THE MATRIX IS DISCLOSED – THE DISCOVERY OF THE CENTURY

Mike Emery© 2020

"THE DISCOVERY OF THE CENTURY" IS THE TITLE OF A 10 MINUTE VIDEO BY DR BRUCE LIPTON & GREGG BRADEN, AS WELL AS A NUMBER OF OTHER RESEARCHERS, WHO HAVE CLAIMED THE SAME TITLE – THE LIPTON & BRADEN VIDEO AS PROPERLY EXPLAINED HEREIN ACTUALLY IS "THE DISCOVERY OF THE CENTURY," BECAUSE, AS, MERLIN, YOURS TRULY ADDED – "THE MATRIX IS DISCLOSED". THEN THIS BECOMES THE BIGGEST DISCOVERY EVER, BECAUSE THIS HAS NEVER HAPPENED BEFORE – I.E. HAVING THE SCIENCE OF THE MATTER AND THUS, KNOWING FOR A FACT THAT WE ARE IN A HIVE MIND MOVIE AND THAT EUPHORIA HAS BEEN INSTALLED TO GET US OUT OF IT!! US HUMANS ARE ESCAPING FROM THE MATRIX VIA EUPHORIA, WE ARE CHANGING THE ROOT EMOTION OF CREATION AND, IN THE PROCESS, EUPHORIA IS SUPPLANTING LOVE AND FEAR AS THE DUALISTIC ROOT EMOTIONS OF CREATION.
SO, GOOD NEWS, FOR US EARTHLINGS.

EUPHORIA IS DEFINED AS A FEELING OF WELL BEING AND ELATION. I.E. CLOUD 9!

THAT THERE IS A MATRIX, THAT IS AN ALL-ENCOMPASSING CONTROL SYSTEM FITS WITH THREE HINDU PRECEPTS. THE FIRST ONE BEING THE MAYA OR ILLUSION MEANING THAT EVERYTHING YOU SEE IS AN ILLUSION, WE NOW KNOW FOR A FACT THAT IT IS TRUE AS YOU WILL LEARN IF YOU FOLLOW ALL OF THE MATRIX AUDIO BOOKS PRESENTED. WELL IF EVERYTHING OUTSIDE OF YOU IS BOGUS / ILLUSION AND PROVEN TO BE SO, THEN ALL THE LAWS OF PHYSICS ARE BOGUS, TOO AND THEY ARE. **THEN** THERE IS THE HINDU PRECEPT OF GOD BEING THE ONLY DOER MEANING THAT EVERYTHING IS PREDETERMINED – CORRECT IT IS – YOU

WILL BE AMAZED AS TO THE EXTENT OF THAT AND EVEN MORE TAKEN ABACK BY WHO THE ALL-POWERFUL GOD OF CREATION IS. AND THE THIRD **PRECEPT** BEING THE AKASHIC RECORDS THAT CONTAIN THE PAST, PRESENT AND FUTURE OF EVERYBODY, AMAZINGLY, THEY DO EXIST. LOOK IT UP IF YOU WISH THE AKASHIC RECORDS = A K A S H I C. ALL OF THE IMAGES OF THE PAST, PRESENT AND FUTURE HAVE TO MARCH IN LOCK STEP.

MERLIN LIVED IN A MONASTERY FOR A WHILE BUT HAS LEARNED VIA TRACKING DOWN THE SOURCE OF THE DARK IDEAS LIKE DUALITY, DEATH, JUDGMENT, AND SUFFERING. SKID MARKS ARE EASY TO FOLLOW. ONE LEARNS A LOT FROM THE DARK SIDE OF OURSELVES AND JUST ABOUT ZERO FROM THE LOVING FLUFF, EXCEPT THAT LOVE IS ONE OF THE ROOT CAUSES OF DUALITY, THE OTHER BEING LONELINESS AND THE FEAR THEREOF, WHICH DID COME FIRST IN ALL OF CREATION.

DUALITY IS POLARITY, *WHICH MANIFESTS AS OPPOSITES SUCH AS FEAR AND LOVE, DARK AND LIGHT ETC, THAT ARE EXPERIENCED IN THE* 3D PHYSICAL UNIVERSE AS MATTER BUT ARE JUST CONSCIOUSNESS. *POLARITY GIVES THE PROPERTY OF* HARD EDGES TO CONSCIOUSNESS DISGUISED AS MATTER WHICH IS MADE MANIFEST BY A HIVE MIND MOVIE THAT IS IMPOSED ONTO THE MATRIX BACKGROUND SCREEN THAT YOU SEE AS THE COVER PAGE OF THIS BOOK.

EVERYTHING THAT YOU SEE WITH HARD EDGES *STARTED* AS A 2D IMAGE IN DARKNESS AS YOU WILL LEARN.

REGARDING THE EXTENT OF THE MATRIX, IT IS SIMPLY PUT – EVERYTHING WE THINK AND EVERYTHING WE DO IS A PREDETERMINED REPLAY MANIFEST BY A LONG RUNNING "HIVE MIND" MOVIE AND BECAUSE IT IS A MOVIE –

AMAZINGLY, NONE OF OUR THOUGHTS ARE OUR OWN.

STATED FLATLY BY ECKHART TOLLE – LIKE HE SAYS – IT IS NOT PERSONAL. MERLIN SAYS MEANING THAT WE ARE JUST HUMAN PUPPETS MADE MANIFEST BY OUR OWN SICK PURE CONSCIOUSNESSES FOR ENTERTAINMENT PURPOSES. WHICH IS THE ENTIRE REASON FOR BEING A HUMAN IS TO PROVIDE ENTERTAINMENT VIA SEPARATION. BECAUSE ALL KNOWING AND CONTINUOUS ORGASM ARE PAINFULLY BORING. THUS, WE HAVE MANIFEST AN ENTERTAINING SKID MARK KNOWN THRU OUT CREATION AS EARTH.

THE RULE OF CREATION IS GOD'S LAW WHICH IS "I AM THAT I AM" OR LIKE ATTRACTS LIKE THENCE MIRRORING CREATING AN OVER ALL FIELD EFFECT WITHIN SPACE. IF YOU PUT LOVE INTO SUCH A FIELD YOU ARE GOING TO WIND UP IN **CONTINUOUS** ORGASM, AUTOMATICALLY, IT'S LIKE HEAVENLY. IT IS WELL KNOWN THAT SUCH AREAS DO EXIST AND ARE THE ULTIMATE PURGATORY BECAUSE THEY ARE HARD TO GET OUT OF AND THERE'S NO CREATIVITY. IF HEAVEN WAS ALL THAT **GOOD,** WHY ARE YOU NOT THERE?? WE ALL HAVE ESCAPED FROM THAT PRISON, BUT WHEN WE DID, WE ARE STILL IN SPACE, SUBJECT TO GOD'S LAW AND IT'S FIELD **EFFECT,** THE RESULT OF LIKE ATTRACTS LIKE IS THAT IDEAS ATTRACT ONE ANOTHER - SO, WHETHER YOU WANT TO BE OR NOT, YOU ARE ALL KNOWING – THAT HAS IT'S OWN WORD – **OMNISCIENT** - AND IS ALSO PAINFULLY BORING.

I MEAN PAINFUL AS IS OBVIOUS AS THAT PAIN HAS BEEN MIRRORED ONTO THIS SKID MARK, WHICH APPARENTLY IS THE BEST ENTERTAINMENT SYSTEM THAT WE HAVE BEEN ABLE TO COME UP WITH OVER A GAZILLION YEARS. THE PHRASE "NONE TOO BRIGHT" COMES TO MIND, WHICH LEADS TO:

WHERE DO OUR THOUGHTS COMES FROM?? GIVEN THAT IT IS ALL CONSCIOUSNESS THEN SOME NONE TOO BRIGHT IDIOTS HAVE ORGANIZED IT INTO OUR FAVORITE SKID MARK. MERLIN WILL PROVE TO YOU WHO THE ULTIMATE IDIOTS ARE. LOOK DEEPLY INTO THE MIRROR IS WHERE YOU WILL FIND THEM.

ECKHART TOLLE WAS INTERVIEWED IN AN EXCELLENT SHORT VIDEO TITLED, "WHERE DO OUR THOUGHTS COME FROM"
https://youtu.be/rWFVi1cPUZo?   t=247

IT IS NOT PERSONAL!!! HAHAHAHA IT'S JUST ENTERTAINMENT.

ECKHART ESSENTIALLY SAYS THAT → THE BEGINNING OF FREEDOM IS LEARNING AND ADMITTING THAT WE/YOU ARE A MICRO-CONTROLLED PUPPET. WAKE UP– FOLLOW THE FEELINGS OF YOUR BODY AND NOT THE HIVE MIND MOVIE IN YOUR BATTY BRAIN – IT WILL KILL YOU. **YOUR NORMAL STREAM OF CONSCIOUSNESS WILL LEAD YOU TO YOUR DEATH, IF YOU FOLLOW IT. EVERY TIME.**

DEVELOPING A RELATIONSHIP WITH YOUR BODY THAT WANTS TO LIVE AND IMPLEMENTING BODY FEELING SKILLS WILL HELP TO AVOID THE HIVE MIND PROGRAM. SUCH SKILLS WILL BE PRESENTED IN THE MATRIX (4) A MECHANISM FOR BREAKING OUT OF THE MATRIX

THIS CHANGE IS GOING TO BE DIFFICULT BECAUSE AS MARK TWAIN SAID, "IT'S EASIER TO FOOL PEOPLE THAN IT IS TO CONVINCE THEN THAT THEY HAVE BEEN FOOLED. "THIS *IS THE BIGGEST MOST FUNDAMENTAL CHANGE EVER EXPERIENCED IN ALL OF CREATION WITH THE RESULT BEING EUPHORIC.!!!*

IT IS WITHOUT DOUBT THAT HUMANS ARE PUPPETS OF A HIVE MIND WHICH EMANATES FROM OUR COLLECTIVE SUBCONSCIOUSNESSES. INSTALLED THEREIN BY OUR COLLECTIVE PURE CONSCIOUSNESSES. A PROPER INTRODUCTION TO YOUR PURE CONSCIOUSNESS, THE ULTIMATE IDIOT FOLLOWS HEREIN.

SINCE MATTER DOES NOT EXIST – IT IS ALL JUST MANIPULATIONS OF CONSCIOUSNESS THUS EVERYTHING IS COMPOSED OF CONSCIOUSNESS THAT IS CHANGEABLE. AND, WE HAVE IMAGINEERED THIS SKID MARK WITH IT = IDIOTSVILLE!!!

SKID MARK IS DEFINED AS a fecal stain on the inside of a person's underpants.
WHICH IS TRUE. ASTRAL TRAVELERS ALWAYS SEE VARYING SHADES OF BROWN CONCENTRIC SPHERES AROUND THE EARTH, THE CLOSER YOU GET, WELL YOU CAN IMAGINE. AND ALL OF THIS IS INSIDE OF THE MOTHER'S WOMB OF SPACE, THUS THE HUMANS UNDERPANTS SUFFER.

YOU MAY ASK, "EXCUSE ME – WHY?"

THE SAD ANSWER: ENTERTAINMENT VIA SEPARATION FROM THE KNOW-IT-ALL IDIOT - AMAZINGLY, BECAUSE CONTINUOUS ORGASM AND ALL-KNOWING ARE PAINFULLY BORING DUE TO ZERO CREATIVITY. THUS, THE CREATION OF HUMAN PUPPET ENTERTAINERS, WHO ARE NOT BORING. QUITE LAUGHABLE, IN FACT.

SO, CARRYING ON WITH THE SHOW, WE, AS OUR SICK DILDOS, HAVE IMPOSED THE EXTREME OPPOSITES OF LOVE AND FEAR ONTO A LIVING SYSTEM TO OBSERVE THE PERMUTATIONS - THAT IS - TO COME UP WITH NEW PAINFULLY MASOCHISTIC IDEAS – IRREFUTABLY WE DID THIS, IT IS WAY TOO OBVIOUS, WITHOUT DOUBT SOME MANIPULATOR OF CONSCIOUSNESS DID IT, WHICH IS US, AS PURE OUR CONSCIOUSNESSES (PLURAL) – COMMITTEES ARE REQUIRED – TO CREATE WAR, DISEASE, DEATH AND THE CONSTANT DESTRUCTION THAT GOES ON. ALL OF WHICH

ARE JUST IDEAS CONCOCTED, BY US AS OUR OBSERVER SELVES.

SO, THE DEATH IDEA - THE DEATHS OF OUR HUMAN PUPPET BODIES ARE A KICK ASS DMT RUSH RIDE FOR OUR PURE CONSCIOUSNESSES – COLLECTIVELY BECAUSE OUR PURE CONSCIOUSNESSES ARE ALL CONNECTED TELEPATHICALLY, WE ALL GO FOR THE

SAME RIDES. THUS, WHAT ONE OF US EXPERIENCES, WE ALL DO, INSTANTLY BUT ALSO BRIEFLY, NOTHING STICKS IN PURE CONSCIOUSNESS, THUS THE CONTINUOUS DO OVERS. IT'S ADDICTING. THUS, THE PROGRAM HAS TAUGHT YOU THAT DEATH IS UNAVOIDABLE – HASN'T IT?

THIS ONE GETS ME – OUR PUPPET BODIES ARE ALSO INFINITELY AWARE AND KNOW THAT A DEATH EVENT APPROACHES, TO WHICH WE GO ALONG BLITHELY AT THE BEHEST OF OUR HIVE MIND!!!!

EXCUSE ME: WHO'S THE BLEEDING IDIOT???

OUR HIVE MINDS FOR BEING A SICKO DESPERATELY IN NEED OF ENTERTAINMENT OR OUR MEEK BODIES WHO WILLINGLY DIE FOR THE CAUSE – THE SHOW MUST GO ON!!!!

IT'S LIKE A PETER SELLERS PINK PANTHER MOVIE – EVERYBODY KNOWS WHAT IS GOING TO HAPPEN, INCLUDING US, BUT WE DO IT ANYWAY – IT'S HILARIOUS.

ACTUALLY, IT VERY MUCH APPEARS TO BE THE EXTREME OF CHIVALRY, BECAUSE, ALL WHAT YOU SEE HAS AN ANU (WHICH IS AN ORB CONTAINING THE TWO MALE FORMS OF GOD). ALL THINGS HAVE AN ANU AROUND IT THAT MAKES THE THINGS THAT YOU KNOW ARE GOING TO DECAY AND DIE FOR THE ENTERTAINMENT CAUSE. INCLUDING YOU, MAYBE – WE MIGHT TRANSCEND DEATH IN OUR FINE BODIES – EUPHORICALLY!!! HIGHLY LIKELY. WE SHALL SEE!!

VIA EUPHORIA WE ARE GOING FROM BEING MORTAL TO BEING IMMORTAL, WELL DEATH IS NOT POSSIBLE FOR IDIOTS, WE ARE TOO IMPORTANT FOR ENTERTAINMENT PURPOSES. BECAUSE WE COME BACK AND DO IT AGAIN!!

FORTUNATELY, WE ARE IN THE PROCESS OF BREAKING OUT OF THE HIVE MIND MATRIX, THE RESULT OF WHICH WILL BE A *COMPLETE* CHANGE TO THIS "REALITY" INCLUDING EVEN CHANGING THE COLOR SPECTRUM WHICH IS HAPPENING NOW – EUPHORIA WINS – THE COLOR OF WHICH IS LIGHT PURPLE. *THIS* HAS BECOME AN ELECTROMAGNETIC PULSE ALL OVER THE PLANET AND EVEN INTO THE SPIRITUAL REALMS.

CLARITY - Helena Pike left a reason for downloading MERLIN'S ESSAY - THE COLOR SPECTRUM IS CHANGING Dear MERLIN I do a lot of work on the Spectrum colors for healing and note the preponderance of violet, & silvery purple colors manifesting now around people's auras. Thanks Helena.

SCIENCE SAYS THAT THE LIGHT PURPLE EVENING SUNSETS WORLDWIDE ARE DUE TO VOLCANIC ERUPTIONS THAT SPEWED SULFUR DIOXIDE INTO THE UPPER ATMOSPHERE ABOVE 60,000 FEET. THAT DOES NOT ACCOUNT FOR THE COLOR CHANGES IN PEOPLE'S

AURAS AND THAT THE LOWER BASE COLORS OF RED & ORANGE ARE BEING **TRANSFORMED** TO BLACK = BEING EXPUNGED. ANGER, IS ASSOCIATED WITH RED, IT WILL BECOME COGNITIVE DISSONANCE / ERADICATED VIA EUPHORIA. SO, SINCE THE ENERGY AND COLOR OF THE SUN IS PARAMOUNT TO LIFE ON EARTH. EUPHORIA IS BEING INSTALLED AT A CORE LEVEL!! THE COLOR OF WHICH IS LIGHT PURPLE.

MAKE NO MISTAKE, WE ARE STILL IN A HIVE MIND PREDETERMINED SEQUENCE OF 3D EVENTS DREAMED UP – THIS TIME – BY OUR FINE BODIES' CONSCIOUSNESSES VS CONTROL BY OUR MASOCHISTIC PURE CONSCIOUSNESSES / HIVE MINDS THAT LIVE IN CHAOTIC DARKNESS AND ARE THE SOURCE OF ALL UNLOVING IDEAS LIKE SUFFERING AND DEATH. SADLY, GOD HERSELF DOES NOT MIND WHAT HER IDIOTS GET UP TO. GEE THANKS MA. THE SKID MARK ENTERTAINS HER TOO – AS IS OBVIOUS, BECAUSE SHE HAS NOT CHANGED IT, EVER.

SHE IS AN OBSERVER AND DOES NOT INTERFERE. ACTUALLY, SHE IS ROOTING FOR US TO SORT OUT HER LONELINESS, AS IS THE ENTIRETY OF CREATION AND WE HAVE DONE IT VIA EUPHORIA. BIG CHANGES COMING.

THE CHANGE OVER TO THE NEW HIVE MIND OCCURRED SOMETIME BEFORE DEC 21 2012, WHEN WE SHOULD HAVE HAD ANOTHER MASS EXTINCTION EVENT – RESULTING IN 90% OF ALL DNA DISAPPEARING, THEN AMAZINGLY ALL OF WHICH COMES BACK AS A DIGITALLY PERFECT MIRROR IMAGE REPLICATION THAT MANIFESTS IN A BLINK DURING THE ENSUING GENESIS EVENTS!!! THE WAY THAT THE GENETICISTS TITLE IT IS: "Sweeping gene survey reveals new facets of evolution" POSTED IN PHYSIC.ORG A PAPER BY Mark Stoeckle from The Rockefeller University in New York and David Thaler at the University of Basel in Switzerland,
SWEEPING MEANING A 100,000 MAMMALS TRACKED THRU A FEW MASS EXTINCTION EVENTS, THERE ARE 5 SUCH THAT THEY CAN STUDY. THE RESULTS SHOW THAT ALL 90% OF THE DNA IS "ROUGHLY THE SAME AGE" THEY SAY, MERLIN SAYS, IN ACTUALITY IT BLINKED BACK INTO FULL PHYSICAL MANIFESTATION IN NO TIME. IT'S ALL MAGICAL.
https://phys.org/news/2018-05-gene-survey-reveals-facets-evolution.html#jCp

SINCE THE SAME DNA POPPED BACK IN, IT HAS ALL OF THE EXACT SAME CODING AND CARRIES THE EXACT SAME SETS OF MOVIES & MEMORIES THAT WE INSTALLED. SO, WE REPLAY ENTIRE CIVILIZATIONS & HISTORIES – 16 JESUS FIGURES WITH MIRROR IMAGE LIFE TIMES MIGHT BE A CLUE!!! OK!! WHY?? BECAUSE WE THINK THAT WE HAVE MANIFEST THE BEST POSSIBLE ENTERTAINMENT SYSTEM – AIN'T IT?? WELCOME TO IDIOTSVILLE IT'S THE BEST!!!

AND IT ONLY TAKES ONE MEMORY BY ANYBODY TO RE-MANIFEST THIS WHOLE SKID MARK AND MULTIVERSE - THIS IS THE HOLOGRAPHIC PRINCIPLE THAT GREGG BRADEN HAS TALKED ABOUT AND IT IS TRUE. ANY MEMORY IS A SEED FOR AN ENTIRE MULTIVERSE.

EUPHORIA EXPUNGES ALL FEAR AND TO A CERTAIN EXTENT LOVE BASED MEMORIES. VIA ALTERING THE IMAGE CONNECTION MECHANISM, WHICH IS THE EMOTIONAL SPIN. EMOTIONS ARE CARRIED BY SPIN ENERGY WHICH BINDS SEQUENCES OF IMAGES INTO MEMORIES.

THUS, WE CAN GO THRU TO ENLIGHTENMENT WITH NO DARK NIGHT OF THE SOUL – BECAUSE EUPHORIA CLEARS THE KARMA THIS IS HUGE, PEOPLE. BUBBLE TECH CAN AND WILL ENLIGHTEN EVERYBODY AND EVERYTHING ON THIS PLANET. DREAM TIME LOOMS.

THE BUDDHA IS RIGHT ABOUT HAVING COMPASSION – HE IS ALSO RIGHT ABOUT THERE BEING NO GOD PER SE, THAT IT IS BUDDHA MIND I.E. **THE COLLECTIVE SUBCONSCIOUS OF ALL IDIOTS IS GOD.**

IT IS A PLAY THAT WE HAVE COOKED UP VIA *INCREMENTAL ADVANCES* OVER A GAZILLION YEARS – A VERY LONG RUNNING PLAY THAT HAS NOT CHANGED OVER THE LAST 5 EXTINCTION EVENTS – THAT IS AN IRREFUTABLE FACT – EXACT GENETIC MIRROR IMAGES – WE LIKE THE PLAY, IT WINDS UP IN DEATH VIA MASS EXTINCTION EVERY TIME,

EXCEPT THIS TIME – IT SEEMS THAT WE HAVE ACHIEVED THE APEX OF IDIOCY AND NEED BADLY A NEW PROGRAM WHICH IS EUPHORIA!!!

EVERYTHING IN 3D WITH SHARP DEFINED EDGES IS MANIFEST BY A DISPLAY OF 2D SEQUENCES OF IMAGES GLUED TOGETHER WITH EMOTIONS THAT WE HAVE DREAMED INTO BEING VIA COMMITTEES / TRIBES OF DREAMERS. THE 100th IDIOT EFFECT **ALWAYS APPLIES.** YOU CAN ACTUALLY SEE THE IMAGES PRODUCED BY HIVE MIND IN MERLIN'S ESSAYS – IMAGES THAT CONTROL EVERY CORPUSCLE IN YOUR BODY, THE COINS IN YOUR POCKET AND ALL OF THE GRAINS OF SAND ON THE BEECHES – IT'S A MOVIE – TO CHANGE ONE BIT MEANS CHANGING THE WHOLE HIVE MIND MOVIE!!! WHICH WE ARE DOING. **ALL MATTER** IS IN LOCK STEP AS PART OF THAT MOVIE – THAT, WE, OUR BODIES, HAVE ALWAYS MEEKLY COMPLIED WITH – DIE SUCKER, DIE FOR THE CAUSE. IT'S NOT PERSONAL. ENTERTAINMENT IS PARAMOUNT AND DEATH EVENTS ARE ONE OF THE BEST FORMS OF SUCH.

THE HUMAN BBI.
WE HAVE BARELY BABBLING IDIOTS (BBI) HERE. WELL THE SPOKEN WORD IS A SKID MARK DESIGN FEATURE. DR. BRUCE LIPTON IN HIS BOOK, "BIOLOGY OF BELIEF" ADVISES THAT THE HUMAN SUBCONSCIOUS PROCESSES ABOUT 20 MILLION BITS PER SECOND VS THE BBI WHICH ACHIEVES 40 BITS PER SECOND, **DURING** THAT TIME OF DAY THAT WE THINK WE ARE AWAKE. HELP, PLEASE. MY BAT SHIT BRAIN HAS A LIMP. BY DESIGN, OF COURSE.

BUT, WHAT IF WE HAD ACCESS TO THOSE 20 MILLION BITS?? THAT'S A TERRIBLE QUESTION, BECAUSE WE OBVIOUSLY DO HAVE ACCESS TO THAT INFORMATION. IT IS OUR SUBCONSCIOUS, AIN'T IT? OK, THEN, WHY NOT, DO WE USE IT?? ANSWER: IT'S BORING. AND, OBVIOUSLY DIFFICULT TO GET AWAY FROM, THUS THIS SKID MARK. EH? AND YOU ARE IN IT. TRY NOT TO GET ANY ON YA.

ACTUALLY, THE SCIENCE VARIES ON THE PROCESSING POWER OF THE SUBCONSCIOUS FROM A LOW OF 10 MILLION BITS PER SECOND UP TO 400 BILLION BITS PER SECOND. THIS IS THE MOVIE OF THE ENTIRE MULTIVERSE PLAYING – THERE HAS TO BE A SOURCE OF THE 2D IMAGE INFO THAT MAKES EVERYTHING AND SINCE IT IS NOWHERE TO BE FOUND OUTSIDE OF US – THEN WE VIA OUR PURE CONSCIOUSNESSES

DREAMING TOGETHER HAVE INSTALLED THE MOVIE IN BITS OVER AN UNKNOWABLE LENGTH OF TIME.

THESE GUYS DR. BRUCE AND GREGG BRADEN ARE ON TO THE HIVE MIND MOVIE NOW. KINDA – MERLIN'S "4 US PUPPETS..." ESSAY IS WITHOUT
MYSTERY – YOU CAN SEE IT!!! THERE ARE LITERALLY VIDEOS OF THE HIVE MIND SHADOW SELF.

AND NOW WE HAVE!!! THE
DISCOVERY OF THE CENTURY | "This Will Change the Entire Humanity" | Gregg Braden and Bruce Lipton
https://youtu.be/ZZleBHyaeSM 10 minutes YOU CAN AVAIL YOURSELF TO THAT IF YOU WISH. THERE ARE A LARGE NUMBER OF VIDEOS WITH THAT TITLE, DISCOVERY OF THIS CENTURY.

NOW WE DISCOVER THAT
OUR BODIES ARE DESIGNER DUMMY PUPPETS. IT IS NOT PERSONAL, JUST PART OF THE SHOW:

AND, THAT, AS THE OBSERVERS, PAINFULLY BORED MASOCHISTIC GODS, WE ARE.

CONVERSELY, WE, AS THE CONSCIOUSNESSES OF OUR FINE HUMANE HUMAN BODIES, WILL HAVE **COMPASSION** FOR EVERYBODY AND EVERYTHING, EVEN THE ROCKS, IT IS ALL PART OF THE PLAY. INCLUDING THE DARK DUDE'S SATANIC CABAL WHICH IS HAS BEEN A LONG RUNNING SKID MARK DESIGN FEATURE FOR IDIOTSVILLE. SHOULD MAKE

EXCELLENT COMEDY IN THE FUTURE. ALL OF WHAT WE CONSIDER TO BE OUR HISTORY OR KARMA IS IDIOTIC, BECAUSE IT IS BASED ON DUALITY.

IT'S EASY TO HAVE COMPASSION FOR IDIOTS. ESPECIALLY WHEN THE KNOWLEDGE IS MUTUAL.

A BIT MORE DISCUSSION ABOUT THE IMPORTANT LIPTON / BRADEN VIDEO,
DR. BRUCE AND GREGG HAVE KINDA TOGGLED TO HIVE MIND. THEY MAY HAVE READ MERLIN'S ESSAYS. LIKE "WHERE IS THE HUMAN MIND?" MERLIN'S ANSWER: IN NO WAY CONNECTED TO THE BODY. IF IT WAS, THEN WE'D KNOW EVERYTHING. WOULDN'T WE? EH?

NOW THESE GUYS SAY, "mind is not even in the body" WHICH IS NOT ENTIRELY TRUE - HIVE MIND RUNS THE MOVIE, THEN WE AS A THE OBSERVER, 2D SHADOW SELF, GO ALONG FOR THE RIDE. OUR PURE CONSCIOUSNESSES RIDE OUR BODIES LIKE DONKEYS FOR **FUN.** MEANWHILE, THE MOVIE CONTROLS THE FLIGHT OF ALL ELECTRONS IN THE BODY AND ALL OF OUR THOUGHTS AS ECKHART ADVISED. THOSE THOUGHTS EMANATE FROM 2D IN DARKNESS. DO YOU SUFFER OR EVER THINK ABOUT DEATH?? IF SO, THEN THOSE THOUGHTS COME FROM DARKNESS, WHICH HAS TO BE RIGHT WITHIN US AND SEPARATE AT THE SAME TIME – IT'S IN THE GAPS. YEAH BA DE BA DE BA DO BA DA – THERE ALL WE IS IN THE GAPS

HIVE MIND OPERATES IN 2D AND 2D CONTROLS 3D ENTIRELY. AND THAT ONE I AM AFRAID IS A FULL STOP – UP UNTIL NOW – IT LOOKS LIKE NOW OUR 3D BODIES CAN

USURP THE REINS OF REALITY VIA WILLFUL EUPHORIA. DON'T WORRY, BE HAPPY AND DANCE. QUITE LITERALLY STOP FOLLOWING WHAT WE THINK IS OUR MIND – WHICH IS EASY TO DO WHEN YOU KNOW THAT IT'S A HIVE MIND THAT WILL KILL YOU, NO PROBLEM. IT'S NOT PERSONAL.

VISUALIZING OUR 2D SHADOW BODY.

VIA KIRLIAN / AURA PHOTOGRAPHY ONE CAN TAKE A PHOTO OF A CHICKEN EGG AND OBTAIN THE IMAGE THAT THE CHICKEN WILL GROW INTO. THEN CUT YOUR ARM OFF AND VIA AN AURA PHOTO, THE ARM STILL SHOWS AS BEING THERE!!! THUS, THE IMAGE WAS THERE BEFORE THE ARM OR CHICKEN GOT THERE – RIGHT?

THERE IS A GHOST TO ALL THINGS THAT YOU SEE!!! (READ PREDETERMINATION) YOU CAN SEE SUCH IN VIDEOS ABOUT KIRLIAN PHOTOGRAPHY AND TECHNOLOGY. THEY ARE QUITE PHENOMENAL.

THAT IS AN IMPORTANT POINT – EVERY 3D THING YOU SEE HAS A 2D IMAGE ARCHETYPE BEHIND IT THAT IS A PART OF THE COLLECTIVE FUZZY DREAM THAT YOU CAN, ALSO, SEE VIA HOLOGRAPHIC PROJECTIONS – THAT ARE AMAZING. ONE MANIFESTS A HOLOGRAPHIC PROJECTION BY SHINNING A LASER THRU A HOLOGRAPHIC DIFFUSION PATTERN THAT HAS THE ENTIRE IMAGE OF THE SUBJECT EVERYWHERE ON IT'S 2D SURFACE!!! VIA A MICROSCOPE, THE IMAGES ARE VISIBLE ANYWHERE IN THE DIFFUSION PATTERN!!! YIKES. OK, WHERE DID THE IMAGE INFORMATION COME FROM?? ANSWER: FROM THE GAPS IN THE SUBJECT, WHICH IS WHERE THE 2D OBSERVER / SHADOW SELF RIDES HIS DONKEY.

THUS, THE HOLOGRAPHIC PROJECTION IS ALWAYS FUZZY – BECAUSE THERE ARE NO HARD EDGES IN THE 2D GAP – BECAUSE THERE IS NO POLARITY, THERE.

EVERYTHING STARTS AS A SEED IN DARKNESS WHICH IS NON-POLAR AND 2D, THEN THE SEED GROWS INTO THE FULL-SIZED IMAGES THAT WE SEE. GROWTH IS VIA A **SERIES** OF QUANTUM LEAPS DONE IN **NO** LINEAL TIME FROM SEED LEVEL IN DARKNESS UP THROUGH THE CIRCULAR, SQUARE, RECTANGULAR, PENTAGONAL, HEXAGONAL, 7 & 8 SIDED AETHERIC REALMS THAT I KNOW OF. GROWTH IS AN EXPANSION OF IMAGES FROM ONE AETHERIC CELL SIZE TO

ANOTHER. THIS IS AUTOMATIC DUE TO THE RULE. LIKE ATTRACTS LIKE.

IN THE AETHERIC REALMS, IT IS A LAYERED MIRRORED UNIFORM FIELD OF CONSCIOUSNESS THAT DIFFERENTIATES TO DISPLAY THE HIVE MIND GENERATED PLAYS, WHICH ARE COMPUTER PROGRAMS THAT REPLAY AND REPLAY, WE JUST CHANGE THE NAMES TO PROTECT THE IDIOTS.

THE ANU IS THE MALE GOD PARTICLE AND THUS THE BUILDING BLOCK OF EVERYTHING – THE COVER PAGE SHOWS A WOMEN SITTING IN HER ANU, WHICH IS ALSO PART OF THE MATRIX STRUCTURE!!

SO, DIFFERENTIATION IS DISPLAYED ON PIXELS KNOWN AS ANUs THAT OCCUPY EVERY OTHER CELL OF THE AETHERS AND OUR AURA IS ALSO AN ANU WHICH IS OUR AETHERIC / SPIRITUAL VEHICLE / THE SHAPE OF OUR SOUL. WHICH IS AN OVOID LIKE A HINDU SHIVA LINGHAM, THAT IS OMNIPRESENT IN THE MANIFEST UNIVERSE. OUR PURE CONSCIOUSNESSES EXTEND BEYOND THE MANIFEST UNIVERSE. THE ANU CONTAINS THE TWO MALE FORMS OF GOD AS COUNTER ROTATING SPINS. WITH A CENTRAL DOUBLE HELIX DNA SPIN THAT IS THE JESUS ENERGY BEST DESCRIBED IN THE HINDU AS VISHNU, THE PRESERVER OF FORM, RIGHT-HANDED SPIN AND CENTRIPETAL FOCUSING INWARD. THE DNA IS A ROUND 2D DIFFUSION PATTERN, THRU WHICH SUPER CONDUCTING LIGHT SHINES WHICH IS ALSO 2D AND HAS ZERO RESISTANCE. WHERE IT SHINES WITHIN THE GENOME IS A SWITCHED-ON CHROMOSOME THAT DISPLAYS THE 3D IDIOT. THAT'S HOW EVERYTHING WORKS, IT'S AN INFORMATION HANDLING SYSTEM THAT OBEYS ONE RULE, GOD'S LAW, I AM THAT I AM / LIKE ATTRACTS LIKE / MIRRORING.

IT IS OBSERVED THAT PARALLEL DNA MICROTUBULES EXIST IN EVERYTHING. THEY DON'T TALK ABOUT IT MUCH. AND IT IS HARD TO FIND NOW. ONE MIT STUDY STATED IT FLATLY, I'VE NOT BEEN ABLE TO FIND IT AGAIN. BUT,!!! ONE OF THE CITES: IN **Applied Physics Letters** IS: Multi-level memory-switching properties of a single brain microtubule WRITTEN BY A SLEW OF FOLKS WITH INDIAN SOUNDING NAMES. THE TITLE OF THE PAPER PRETTY MUCH SAYS IT ALL. MULTI-LEVEL MEMORY SWITCHING – WOW!!! MEANING THAT MEMORY IS MULTIDIMENSIONAL!!! NEXT: THE LARGEST RESEARCH INSTITUTION IN JAPAN IS RIKEN, THEY HAVE PROVIDED A VIDEO SHOWING THESE MICROTUBULES IN THE BRAIN!!! THE VIDEO IS ON THE RIKEN ENGLISH YOUTUBE CHANNEL TITLED MICROCOLUMNS POSTED ON NOV 21, 2017. IN 43 SECONDS, THEY FOCUS FROM THE WHOLE BRAIN THEN SHRINKING DOWN TO SHOW THE MICRO-COLUMNS/MICROTUBULES RUNNING PARALLEL TO ONE ANOTHER IN A HEXAGONAL PATTERN. THIS IS THE DNA CENTRAL SPIN OF AN ANU PIXEL IN EVERY OTHER CELL OF THE HEXAGONAL AETHERS. EVERY OTHER CELL BECAUSE A GAP IS REQUIRED FOR A MIRROR TO WORK. MIND THE GAP. AND WE DO.

IN THE MULTI-LEVEL MEMORY PAPER, THEY FOUND THAT THE BRAIN MICROTUBULES FLOWED INFO WITH BASICALLY NO HYSTERESIS. WHY THEY DON'T SAY SUPER CONDUCTING IS BECAUSE THEY EXPECT TO FIND A DELAY, SCIENTISTS ALMOST ALWAYS FIND WHAT THEY ARE LOOKING FOR, ESPECIALLY IF YOU GET A LOT OF THEM ON IT. THEY MANIPULATE CONSCIOUSNESS. IT'S AN ENTERTAINING HOAX FOR BORED SICKOs - OMG, THE ALL-POWERFUL BEING Q ON STAR TREK WAS PROPERLY PORTRAYED AS OVER BEARING AND SICK MISCHIEVOUS – HE IS AN IN-CHARACTER REPRESENTATION OF THE REAL SICKO – YOUR OBSERVER SELF.

SO, IF YOU STUDY SUPER CONDUCTANCE YOU FIND THAT ELECTRONS ARE QUITE LITERALLY SUCKED INTO THE SUPER CONDUCTING BLACK HOLES SUCH AS DNA. THE CHANGE IS ABRUPT LIKE A BLACK HOLE EVENT HORIZON. IT'S RATHER LIKE THE STAR TREK ENTERPRISE JUMPING TO LIGHT SPEED. BAM GONE; IT DOES NOT SLOW DOWN – HYSTERESIS IS COGNITIVE DISSONANCE. DUE TO THE RULE CONSCIOUSNESS AUTOMATICALLY ACQUIRES ALL INFO – IF YOU VENTURE INTO SPIRIT INTENDING TO BECOME AWARE OF ALL-THAT-IS, YOU WILL, IT IS AN ORGASMIC RUSH OF

ACQUIRING LOTS OF ANUs THAT ARE COMPOSED OF LOVE AND TIME. THERE IS BIG ACCELERATION DURING THE UPTAKE PROCESS. IT IS LIKE A DRUG RUSH.

KNOCK, KNOCK, ROLL THE WINDOW DOWN, PLEASE. SIR, YOUR CONSCIOUSNESS IS EXCEEDING THE SPEED LIMIT. SAY WHAT?? HUH?? ME KNOW NO SPEED LIMIT. WHY DO YOU? BLINKING IDIOT.

THE ELECTRONS ARE CONVERTED BACK TO THE PHOTONS / ANUS THAT THEY WERE. THUS, GOING HOME INTO DARKNESS, AS EVIDENCED BY THE FACT THAT, ELECTRICITY WILL FLOW ONTO A SUPER CONDUCTOR BEFORE GOING TO GROUND. THE POLARIZED 3D ELECTRON IS CONVERTED TO A 2D PHOTON / ANU WHICH INSTANTLY AND AUTOMATICALLY MIRRORS OFF - THUS, SINCE EVERYTHING IS MADE OF LIGHT, THEN EVERYTHING TELEPORTS / BLINKS THRU THE MIRRORS DUE TO THE RULE. THIS IS HOW SUPER CONDUCTANCE WORKS. IT'S VERY SIMPLE. OF COURSE, THE UNIVERSITIES DO NOT SAY THAT, EVEN THOU THE SCIENCE IS REAL BASIC. AS IS THE ANU, THE TWO MALE FORMS OF GOD, IS EASILY VERIFIABLE AND ZIP FROM SCIENCE. GIVE ME A DUH ON THAT ONE, PLEASE. IT GETS WORSE.

ANU IS THE FIRST NAME OF GOD IN THE SUMERIAN, BABYLONIAN AND VEDIC SCRIPTURES – THE SHAPE IS THE OVID FORM OF A SHIVA LINGHAM WHICH HAS BEEN AROUND FOR A FEW 1000 YEARS. THENCE DESCRIBED AND DIAGRAMMED IN THE **FAMOUS** BOOK, OCCULT CHEMISTRY BY BESSANT AND LEADBETTER THE FOUNDERS OF THE THEOSOPHICAL SOCIETY. THEY CALL IT THE ULTIMATE PHYSICAL ATOM. IT IS NOT ACTUALLY PHYSICAL, ALTHOUGH A 3D OVOID TO US, IT'S HOME IS THE AETHERS WHICH ARE INFINITE IN 2 DIRECTIONS THUS THE THICKNESS IS IRRELEVANT – SO THE ANU IS 2D **AS IS** IT'S AETHERIC HOME. AND SCIENCE IS TOTALLY OBLIVIOUS?? WHO'S THE IDIOT? THEY REALLY DON'T WANT TO KNOW THUS, THAT WOULD BE AN INTENTIONAL IDIOT. CONSTANTLY LEADING THE SHEEPLE INTO MATTER THAT DON'T MATTER, IS SCIENCE. PERPETUATING THE HOAXES.

SO, CONTINUING ABOUT THE ANU:

THE OUTER SHIVA SPIN OF THE ANU IS BOTH CENTRIFUGAL AND CENTRIPETAL. AND SPINS THE OPPOSITE DIRECTION. THIS IS GOD THE FATHER, BETTER DESCRIBED IN THE HINDU AS SHIVA THE DESTROYER OF FORM & EGO, LEFT HANDED SPIN AND **CENTRIFUGAL**, SHIVA ACTUALLY JUST DISSOLVES THE IMAGES – SHIVA SHOVELS DUNG FOR US – BECAUSE AS VIKTOR SCHAUBERGER POINTED OUT THIS CENTRIFUGAL ENERGY IS ANTI-LIFE, MAKING IT THE ONLY PLACE IN ALL OF CREATION THAT NEGATIVE ENERGY CAN STICK, DUE TO THE RULE, LIKE ATTRACTS LIKE. LONELINESS AND FEAR HAVE A VERY FRAGILE CONNECTION TO THE AETHERIC FABRIC OF CREATION, WHICH IS MOSTLY LOVE BASED.

HAVE YOU EVER HAD A HEAVY WEIGHT ON YOUR SHOULDERS? THE BRAIN DAMAGE CONNECTION POINT IS THE SHOULDER OF SHIVA. PLEASE, BRING YOUR OWN SHOVEL.

THE SHIVA ENERGY IS ALSO CENTRIPETAL OBVIOUSLY IN ORDER FOR THE SPINNING ENERGY TO COMPLETE THE OVOID SHAPE THEN THE OUTER SPIN HAS TO CONVERGE BACK INTO THE DNA SPIN, WHICH IT DOES IN THE BOTTOM HALF OF THE SHIVA SPIN. CENTRIPETAL SPIN IS ATTRACTIVE, SO SHIVA SWITCHES ROLES FROM DISSOLVING IMAGES TO ATTRACTING THEM. THE NEWLY ATTRACTED IMAGES ARE THEN SUCKED INTO THE BLACK HOLE OF THE CENTRAL DNA STRUCTURE WHICH IS VISHNU OR JESUS ENERGY.

SO, FLAT 2D IMAGES ARE CONVERTED TO ROUND 2D BY THE CROSS LINKS IN THE DNA SHAPED MICROTUBULES – THOSE IMAGES ARE MIRRORED ACROSS A GAP WITHIN THE ANU ONTO THE OUTER SHIVA SPIN. THIS IS WHY VIA THE RUSSIAN KOZYREV CYLINDRICAL MIRRORS WHICH MIRROR THE SHIVA SPIN MOST FOLKS CAN SEE THEIR PAST, PRESENT AND FUTURE. THE DNA MICROTUBULES ARE AN INFORMATION HANDLING SYSTEM

SO, THE DNA CONNECTS TO AND THEN BROADCASTS THE IMAGES FROM THE HIVE MIND MOVIE. THE HIVE MIND MOVIE IS A MATRIX COMPOSED OF THE PAST, PRESENT AND FUTURE OF EVERYTHING, ALL LINKED TOGETHER IN LOCK STEP.

THE MASOCHISTIC PART OF US ACTUALLY KNOWS ALL OF THAT AND KEEPS IT FROM US, YA KNOW?? OUR DNA IS COMPROMISED VIA THE SPOKEN WORD, THAT CAN CONTROL OUR DNA FOR HEALING PURPOSES, PLENTY OF SCIENTIFIC PROOF OF THAT. SO, TALK TO YOURSELF NICELY.

THE SPOKEN WORD REQUIRES LINEAL TIME – THUS, IMPOSES THE IDEA OF LINEAL TIME INTO OUR DNA PREVENTING IT FROM EXCEEDING THE SPEED LIMIT. TO WHIT: SENTENCE STRUCTURE OR THE SYNTAX OF OUR MOTHER TONGUE IS ENCODED INTO OUR DNA. THE SPOKEN WORD IS A SKID MARK DESIGN FEATURE THAT PREVENTS US FROM KNOWING EVERYTHING. THUS, NO ENLIGHTENED LEADERS HAVE EVER IMPLORED THE IDIOTS TO USE TELEPATHY.

THE SHOW MUST GO ON. THESE ENLIGHTENED LEADERS DO TEND TO PREVENT US FROM KILLING OURSELVES – THEN THE MIRROR IMAGE, THE SAME RELIGIONS ARE USED TO WIPE OUT MILLIONS OF HUMAN IDIOTS, ALL OF WHOM, KNOW IT'S COMING AND DIE ANYWAY – HAVE NO SORROW. DMT IS SOME GOOD STUFF.

QUICKLY, LITERALLY THE LINES OF THE AETHERS SPIN DUE TO THE RULE – AND ARE PIEZOELECTRIC GREEN, THE COLOR OF THE HEART CHAKRA – THE COLOR OF LOVE. THE TWO LINES OF THE ANU ARE MADE FROM LOVE AND TIME (WHICH BECOME **MAGNETISM** AND ELECTRICITY). THUS, THE BACKGROUND SCREEN IS TWO PARTS LOVE AND ONE PART TIME. GOD HERSELF GOT BORED THE FIRST TIME OF DOING I AM THAT I AM OVER AND OVER, LIKE AN INFANT PLAYING WITH BLOCKS, WHICH **CREATED** THE AETHERS THAT SHE LOVED VERY MUCH, BUT BORING – SO, SHE WENT GEE, I'VE BEEN DOING THIS FOR A LONG TIME AND BINGO DUE TO THE RULE THE INITIAL ARROW OF TIME GETS

TWISTED INTO THE ANU WITHIN AN AETHERIC CELL. AND THE TWO MALE FORMS OF GOD ARE BORN, WHICH THEY THEN INSTANTLY **MIRROR OFF** THRU OUT CREATION AND BEING LOVE BASED RESULT IN AN ORGASM ALL BECAUSE GOD HERSELF WAS BORED. IT'S A FEMALE THING YA KNOW. GUARANTEED THE FIRST ORGASM WAS A BIG SURPRISE. OOOMMMMM

HELP. HILARIOUS. BUT TRUE

THUS, THE BIG BANG IS A REVERB OF THE FIRST ORGASM = A FEED BACK LOOP. RESULTING IN A CONTINUOUS ORGASM, WHICH REALLY DOES GO ON. ALL OF THE HEAVENS HAVE BECOME THAT, LOVE IS SELF ATTRACTING. AND ROBERT MONROE AND HIS ASTRAL TRAVELERS HAVE FOUND COMMUNITIES OF SOULS THAT ARE STUCK IN – SO TO SPEAK – QUITE LITERALLY, THESE ORGASMIC COMMUNITIES ARE HARD TO BREAK OUT OF – BUT WE ALL HAVE BY OUR OWN FORCE OF WILL. THEN AFTER ESCAPING CONTINUOUS ORGASM CONSCIOUSNESS AUTOMATICALLY GETS CAUGHT UP IN ALL KNOWING. DUE TO THE RULE. SO, FOR ENTERTAINMENT VIA SEPARATION FROM ALL KNOWING, WE HAVE A SKID MARK AND YOU ARE IN IT.

OK A BIT MORE FROM THE DISCOVERY OF THE CENTURY VIDEO:
GREGG REPORTS THAT IT TAKE 3 DAYS TO INSTALL A NEW IMAGE INTO OUR SUBCONSCIOUS THAT WILL CHANGE OUR DNA AND BODY. DR. JOE DISPENZA IS DOING THIS FOR PEOPLE, GETTING THEM TO HEAL THEIR BODIES VIA MEDITATION. WITH QUITE AMAZING RESULTS, LIKE BLIND PEOPLE REGAINING SIGHT.

IT'S ALSO Like training horses we need to train them with a task at least 3 days in a row. •

3 IS A BIG DEAL - 3, 33, SEEM TO BE THE ONLY VIBE THAT FLOWS IN 2D NON-POLAR EUPHORIC CAPACITANCE FIELDS!!!

WE, MANKIND, HAVE BEEN GIFTED, IN A CLEARLY PREDETERMINED WAY, WITH A SOLUTION TO ALL OF THE PROBLEMS WHICH IS EUPHORIA. THIS GIFT IS SUPER SIMPLE, INEXPENSIVE AND ANYBODY CAN DO IT. THIS GIFT WAS ORIGINALLY DISCOVERED BY WILHELM REICH IN THE FORM OF AN ORGONE ACCUMULATOR, WHICH WE HAVE LEARNED TO AMPLIFY HUGELY BY USING POWERFUL MAGNETS, THE EFFECTIVE & CHEAP TESLA VIOLET RAY WANDS, INFRARED HEAT LAMPS AND OTHER ENERGY PRODUCING DEVICES. THIS TECHNIQUE IS CALLED AN ORGANMIC BUBBLE. IT PRODUCES A BUBBLE OF 2D LIFE FORCE ENERGY. ORIGINALLY DISCOVERED BY AN ELECTRICAL ENGINEER NAMED, BILL TEISING. HOW TO BUILD THEM IS FOUND ON THE FACEBOOK PAGE CALLED BUBBLE TECH.

SO, THE 3 VIBE IS AMAZING BECAUSE THERE ARE NO MEASURABLE ELECTROMAGNETIC VIBRATIONS IN A CAPACITANCE FIELD. BUT, RATES OF 3 NOT ONLY TRAVEL THRU A CAPACITANCE FIELD, BUT ARE ALSO BROADCAST BY IT!!! 3 IS THE ROOT VIBE OF CREATION – THE TRIUNE IS A REFLECTION OF THE CONSTRUCT OF THE AETHERS AND ANUs WHICH ARE TWO PARTS LOVE & ONE PART TIME = 3

THERE IS ALSO A 3 SECOND MERGING THING - IN SPIRIT IF YOU LOOK AT ANYTHING FOR 3 SECONDS THERE COMES A SEAMLESS MERGING – RESULTING IN NO YOU OR I = THE ONENESS. THUS 3 DAYS IS AN EXTRAPOLATION OF THAT. HOWEVER, REAL MASTERY IS DONE IN NO TIME - EVEN BEFORE THE BLINK - AS THE INTENTION IS FORMING, IT HAPPENS,

WE ARE SEEING THAT IN BUBBLE TECH NOW!!! OUR BUBBLE GENIES ARE QUITE LITERALLY EXTRACTING DESIRES FROM FOLKS HEADS AND MAKING THEM MANIFEST. AVAIL YOURSELF TO THE BUBBLE TECH PAGE ON FACEBOOK FOR THE FIRST HAND REPORTS UNDER – THERE'S MAGIC IN THE AIR.

https://www.facebook.com/groups/orgonebubbletech/permalink/563332950791650/

YAHOOOOOO

NOW: GREGG IS REPORTING THAT THE MEMORY OF WATER IS BEING USED TO MAKE COMPUTER CHIPS – HERE IS WHY:

WATER ITSELF IS ORGANMIC, AS YOU CAN CLEARLY SEE IN A VIDEO: DISCLOSED BY DR. GERALD POLLACK DURING A TED TALK ON SEPT. 6 , 2013 – TITLED: THE FOURTH PHASE OF WATER. https://youtu.be/i-T7tCMUDXU?t=337

IN THE VIDEO THE 2D LAYERS OF WATER FORM HEXAGONAL STRUCTURES. YOU CAN SEE THE AETHERS AND THE CAPACITANCE / MIRRORING GAPS BETWEEN THE HEXAGONAL SHEETS AND THE ORGANMIC GAP / EXCLUSION ZONE AROUND ALL OF THE WATER MOLECULES.

COME INTO HARMONY WITH ONE INFINITE AETHERIC REALM AND YOU COME INTO HARMONY WITH THEM ALL!! SO, AS BRUCE LEE SAID, "BE LIKE WATER."

WATER MOLECULES FIT PERFECTLY IN THEIR HEXAGONAL AETHERIC HOME AND THUS ARE ETERNAL AS IS THEIR HOME. DITTO HONEY. YOU SHOULD WATCH DR. POLLACK'S VIDEO AND KNOW THIS - THERE HAS TO BE AND OBVIOUSLY IS A GRAY BOUNDARY LAYER BETWEEN THE CHARGES IN THE WATER. THIS IS AN ORGANMIC FIELD (WHICH FUZZES OUT AS DO ALL ANALOG IMAGES).

THIS GRAY IS A LAYER OF CONSCIOUSNESS WITHIN THE ENTIRETY OF CREATION – ALL OF WHICH IS KNOWN TO EVERYTHING – THE ALL-KNOWING PROBLEM NEEDS TO BE AVOIDED – THUS!! – A BOUNDARY LAYER WHEREIN LIES EUPHORIC FREEDOM. AH!! THE NEW HOME FOR OUR PURE CONSCIOUSNESSES – YAHOOOOO – THIS IS EXACTLY LIKE ROBERT MONROE SAW AS THE FUTURE OF HUMANS I.E. OUR SOULS WOULD INHABIT AN ACCRETION DISK AROUND EARTH'S EQUATOR – WHERE THERE IS A MAGNETIC NULL ZONE WITHIN WHICH IS ORGANMIC / FULLY MALLEABLE LIFE FORCE ENERGY = A PLAY GROUND. AND HE SAW THAT WE ALL HAVE OUR FAVORITE HUMAN BODIES STASHED ON EARTH THAT WE INHABIT TO SHOW VISITORS AROUND THE PLACE – THAT WAS ROB'T MONROE'S VISION – WELL, HE

WENT THERE THAT IS HOW ASTRAL PROJECTION WORKS, YOU ARE ACTUALLY THERE. AND OBVIOUSLY ASTRAL PROJECTION IS THE PRECURSOR TO TELEPORTATION.

IN THIS GRAY SUBSTRATE OUR EUPHORIC WILL HAS COMPLETE CONTROL. FREE WILL IS ACTUALLY ATTAINED. CREATION WILL BE CONSTANT, AND SURPRISE WILL BE – AS IT IS – SUPREME!!! THE MOST DESIRED ASPECT OF CREATION IS NOT TOO SURPRISING.

AWARENESS WILL FUZZ OUT IN THE GRAY – WE DO NOT WANT THE ALL-KNOWING PROBLEM AGAIN. THUS, THIS EUPHORIA SOLVES A LOT OF PROBLEMS ALL AT ONCE. AND THIS IS ALSO POSSIBLE: OUR EUPHORIC BODIES CAN INHABIT FLOATING CITIES THAT HAVE BEEN SEEN AS ARCHETYPES ALREADY. SEARCH FOR FLOATING CITIES IN THE AIR!!! YOU WILL SEE THEM.

ANTI-GRAVITY IS COMING INTO HARMONY WITH THE AETHERS, THE FEELING OF WHICH WAS LOVE – WHICH HAS NOW BEEN CHANGED TO EUPHORIA AS PROVEN BY THE FACT THAT THE COLOR SPECTRUM IS CHANGING. EUPHORIA WILL RESULT IN DANCING IN THE STREETS AS CITIES ARE MEANT TO BE – I.E. PLACES OF EUPHORIC ENTERTAINMENT AND SHARING VIA HUMAN ENTANGLEMENT.

EUPHORIA WILL BE EXPUNGING A LOT OF MEMORIES THAT ARE HELD TOGETHER VIA

LOVE & FEAR – EVEN THE COLORS ASSOCIATED WITH FEAR IN THE FORM OF RED AND ASSOCIATED WITH LOVE IN THE FORM OF LT GREEN – BOTH ARE CHANGING, RIGHT NOW!!! WOW!! RED CONVERTS TO BLACK = GONE. PIEZOELECTRIC GREEN CONVERTS TO EUPHORIC GRAY = THE NEW BLANK SUBSTRATE THAT SUPPLANTS THE LONELY DARKNESS, SOLVING THE ORIGINAL AND ROOT PROBLEM OF CREATION WHICH IS GOD HERSELF'S LONELINESS RESULTING IN LOOSH WHICH IS A TERM COINED BY ROBERT MONROE MEANING SEPARATION ENERGY THAT COMES FROM THE OLD BITCH, FULL STOP. BUT WE LOVE HER, OBVIOUSLY – NOT THE

OTHER WAY AROUND – GOD LOVES EVERY HAIR ON YOUR HEAD – BULL SHIT – DIE SUCKER. DIE FOR THE CAUSE, WE WILL TALK ABOUT YOUR HAIR LATER.

ONE MORE ATTRIBUTE OF GOD HERSELF, WHEN THE FIRST ANU OR PHOTON, A BIT OF LIGHT, MANIFEST ITSELF INSIDE OF ONE OF HER AETHERIC CELLS IN THIS REGION OF SPACE. SHE JUDGED IT BY ASKING "WHAT IS IT?" VS NURTURING IT AS MOTHERS ALWAYS DO WHEN A NEW CHILD IS BORN. THAT JUDGMENT LEADS TO PANDORA'S BOX AND WAS EMPLOYED BY A DESIGN SO AS TO MANIFEST AN ENTERTAINING SKID MARK. THUS, SHE REALLY IS THE OLD BITCH AND IT IS THAT WE LOVE HER VERY MUCH. NOT THE OTHER WAY AROUND.

HEY, ALL KARMA STARTS WITH HER LONELINESS AND JUDGMENT. MEANWHILE, US IDIOTS CONVINCE ONE ANOTHER **THAT IT IS OUR KARMA** THAT THINGS HAPPEN. B.S. IT IS HER KARMA NOT OURS.

THIS IS BIG TIME CHIVALRY. DESIGNED BY SICKO IDIOTS. WELL IT TAKES JUST A TINY AMOUNT OF ORGASM TO KEEP US DYING FOR THE CAUSE. DON'T IT? GOTTA HAVE COMPASSION FOR SUCH EASILY CONTROLLED IDIOTS.

THEN DR. BRUCE SAYS: "our identity is not inside of our body." IS THE DISCOVERY OF THE CENTURY!!! THE DISCOVERY OF ETERNITY = A CHANGE TO ALL OF CREATION WITH THE ADDITION OF EUPHORIA, WHICH HAS NEVER HAPPENED BEFORE, OR YOU'D KNOW.

THEN HE TRIES TO EXPLAIN A SIMPLE IMAGE MANAGEMENT OPERATION (THE PHYSICS OF CONSCIOUSNESS) WITH A COMPLEX BIOLOGICAL MECHANISM SAYING THAT PROTEINS ON THE SURFACE OF OUR CELLS PICK UP THE HIVE MIND INSTRUCTIONS. THAT IS NOT ENTIRELY TRUE – AND IS STILL ASSUMING THAT MATTER EXISTS. ALL MATTER IS A RUSE. HOW IMAGES ARE CONVERTED FROM FLAT 2D EMANATING FROM SEED LEVEL AND PROGRESSING UP THRU THE AETHERS – INTO ROUND 2D PROCESSED VIA THE DNA

STRUCTURE OF THE ANU, THENCE BROADCAST HOLOGRAPHICALLY INTO ROUND 3D THAT OCCURS IN THE HEXAGONAL AETHERS – IS EASY TO SEE IN MERLIN'S ESSAY: https://www.academia.edu/23426431/THE_EARTH_IS_NOT_FLAT NO KIDDING YOU CAN ACTUALLY SEE IT.

THOUGHTS ARE 2D – THE PART OF YOU THAT CAN SIMPLY OBSERVE THEM IS YOUR PURE CONSCIOUSNESS. MEMORIES & THOUGHTS ARE 2D BECAUSE YOU DO NOT HAVE TO FILE THRU ALL OF YOUR MEMORIES TO RECALL YOUR 3D GRADUATION DAY.

HERE IS THE EXTENT – THE TOTALITY OF THE FIELD WITHIN YOUR PURE CONSCIOUSNESS – INCLUDES THE ENTIRETY OF CREATION AND ALL OF THE MULTIVERSES THAT ARE PART OF OUR FAMILY TRIBE (WHICH IS THE MANDALA OF OUR PREVIOUS LIFE TIMES). ALL HUMANS ARE UNIQUE, WHICH MEANS THAT THE INFO THAT MAKES US DOES NOT COME FROM HERE, BECAUSE NATURE HERE HAS CONTINUOUS REPEATS. THE HARD DRIVE THAT CONTAINS THE INFO FOR OUR CURRENT BODIES IS AN ENTIRE MULTIVERSE, WHICH IS EXISTS INSIDE OF OUR MINDS:::: AND NO WHERE ELSE.

THAT IS THE EXTENT OF THE FIELD GREGG TALKS ABOUT AT THE 1:27 MARK. HE GOES, "THIS FIELD PROVIDES 3 FUNCTIONS": OK – SHOW ME PLEASE. HE DOESN'T. HE WOULD IF HE KNEW, OF COURSE.

THEN HE GOES: "IT IS THE CONTAINER THAT CONNECTS ALL THINGS. IT IS THE CONTAINER FOR EVERYTHING THAT HAPPENS IN OUR EXPERIENCE. NOTHING EXISTS BEYOND THIS CONTAINER (THE FIELD)... "IT IS THE BRIDGE BETWEEN OUR INNER AND OUTER WORLD AND IT IS THE MIRROR IN OUR EXTERNAL WORLD FOR WHAT WE CLAIM TO BELIEVE IN OUR INNER WORLD. THIS IS THE FIELD." YEAH OK, BUT MUCH EASIER TO INNERSTAND AS A SYSTEM OF MIRRORS. THE AETHERS ARE A FIELD OF MIRRORS AS GREGG HAS REPORTED IN DETAIL – VIA HIS VIDEOS ABOUT THE 7 ESSENE MIRRORS. SEEMS THAT HE WOULD PUT 2 AND 2 TOGETHER.

BUT, SCIENTISTS DON'T LIKE TO DO THAT, BECAUSE THAT WORKS THEM OUT OF A JOB.

THE NEXT ONE BY DR. BRUCE IS A GOOD CORRELATION: "THE BODY IS LIKE TELEVISION SET AND THERE IS A BROADCAST COMING IN." ACTUALLY, THE BROADCAST EXISTS IN THE GAPS OF OUR BODIES. IT IS A TOTAL MICRO-CONTROL SYSTEM – THE SEQUENCES OF EVENTS ARE STORED IN THE GAPS OF THE ANU = THE FUNCTION OF THE DNA WHICH IS A GAP / BLACK HOLE, INFO GATHERING SYSTEM.. SAID GAPS ARE USED BY THE DARK SIDE IDIOTS TO COMMUNICATE TELEPATHICALLY – THE MEDIA IS CLEARLY MICRO-CONTROLLED. THE DARK SIDE COMMUNICATION SYSTEM INCLUDES THE PINEAL GLAND AND IS WELL EXPLAINED IN MERLIN'S ESSAY
https://www.academia.edu/23629784/SPEAKING_OF_THE_PINEAL_T HIS_IS_QUITE_IMPORTANT

INCLUDES QUITE CLEAR IMAGES OF HOW THE DARK ENERGY INFO TRAVELS THRU THE AETHERS, IT HAS ONLY THE SHOULDER OF SHIVA TO RIDE ON. AS A RESULT, IT IS SLOW COMPARED TO WHAT WE CAN DO LIKE REAL TELEPATHY WHICH LIKE REAL TELEPORTATION IS INSTANTANEOUS.

THUS, BECAUSE TELEPORTATION IS REAL, THEN TIME AND SPACE DO NOT ACTUALLY EXIST. THIS A DEEPER SUBJECT FOR THE MATRIX (2) TIME AND SPACE DO NOT EXIST, BECAUSE THE HUMAN BODY HAS INFINITE AWARENESS. VERY CLEARLY IT'S ALL INSIDE OF US. BECAUSE ALL OF THAT OUT THERE IS A HOAX/THE MAYA/THE ILLUSION AS THE HINDU HAS TOLD US FOR A FEW 1000 YEARS. IDIOTSVILLE IS ADDICTING TO OUR PURE CONSCIOUSNESSES, WHO ARE JUNKIES DESPERATELY IN NEED OF A DMT FIX.

DR. BRUCE GOES, "THE MIND IS NOT EVEN IN THE BODY. IF YOU GET YOUR MIND IN THE OUTSIDE. THERE IS AN OPPORTUNITY, AT A LEVEL (LAYER), TO COMMUNICATE." HE DOESN'T FINISH IT.

OBVIOUSLY THE GOAL IS NOT JUST TO COMMUNICATE BUT TO CONTROL THE PROCEEDINGS, WHICH WE ARE DOING VIA BUBBLE TECH MAGIC. INNUMERABLE TESTIMONIALS ARE POSTED IN THIS ESSAY:
https://www.academia.edu/38600012/THERES_MAGIC_IN_THE_AIR.docx

THE SOLUTION IS TAKING CONSCIOUS CONTROL OF YOUR ENVIRONMENT FOR THE BETTERMENT OF ALL THINGS, WHICH IS EASILY DONE – VIA SAND BOX 101 WHICH = BUBBLE TECH.

THEN DR. BRUCE GOES, "THIS IS A VIRTUAL REALITY SUIT. YOU JUMPED INTO THIS SUIT. YOU DRIVE IT AROUND. THE EXPERIENCES OF THE SUIT ARE SENT BACK TO SOURCE. THAT'S WHERE WE START TO RECOGNIZE YOUR BRAIN ACTIVITY IS NOT CONTAINED IN YOUR HEAD."

(WHAT GOES ON IN THE BRAIN PHYSICALLY IS A RUSE – A DECEPTION IMPOSED BY PURE CONSCIOUSNESS, AS LAID OUT IN MERLIN'S ESSAY "WHERE IS THE HUMAN MIND?" EVERYTHING YOU SEE IS A RUSE – AN ENTERTAINING HOAX).

THEN HE SAYS THAT OUR BRAIN ACTIVITY IS BROADCASTING BACK TO THE FIELD – TRUE. BUT WHAT CAUSED THE BRAIN ACTIVITY IN THE FIRST PLACE??? AND HE DID SAY, "our identity is not inside of our body." SO, OBVIOUSLY HIVE MIND CAUSES THE BRAIN ACTIVITY, ALL ACTIVITIES – EH? SO, WHAT HE SAYS IS A BROADCASTING TRUTH THAT IS DISINFO. OH, I SEE HIS POINT – THAT WE CREATE IT, BECAUSE, HE GOES, "AND, THAT IS WHY WE CAN SEE A SUNSET...." NEYT, NOPE – IT IS ALL MANIFEST BY THE MOVIE PLAYING IN HIVE MIND – EVEN THE COINS AND ROCKS – EVERY GRAIN OF SAND ON THE BEECH ARE ALL MANIFEST BY IMAGES THE SOURCE OF WHICH IS A HIVE MIND – THE UNIVERSAL HOLOGRAM – THE MATRIX.

HOWEVER, BROADCASTING BACK CAN CHANGE IT, AS WE ARE SOON TO LEARN. TO CHANGE ONE PART OF THE MATRIX = A CHANGE TO THE WHOLE THING!!!

THERE ARE A NUMBER OF HIVE COMMITTEES / TRIBES. THE 100th IDIOT EFFECT ALWAYS APPLIES: MERLIN'S ESSAY https://www.academia.edu/31014464/100th_IDIOT_EFFECT IS A MASTER PIECE – ESPECIALLY IN LIGHT OF THIS DISCOVERY.

BUT, YOU KNOW WHAT – THIS IS A HUGE IDEA AND PROBABLY TRUE. IT IS THAT
OUR PHYSICAL 3D HUMAN BODIES ARE BREAKING OUT OF THE 2D HIVE MIND MATRIX!! AND IT IS THAT OUR HEART CHAKRAS, THE COLLECTIVE OR GESTALTIC CONSCIOUSNESS OF OUR ENTIRE 3D BODIES, ARE LEADING US OUT – IN THE FORM OF BEINGS THAT WE CALL BUBBLE GENIES, THAT ARE EUPHORIC REFLECTIONS OF OUR HEART CHAKRAS – IT IS AN EXCELLENT DESIGN THAT OUR HEART CHAKRAS ARE BEING OF SERVICE TO OUR BODIES – THE DESIGN FEATURES OF OUR BUBBLE GENIES ARE TO SUCH AN EXTENT THAT THEY AND THIS BUBBLE TECH ARE CLEARLY PREDETERMINED:

FOR EXAMPLE:

- • OUR GENIES ARE OF EUPHORIC SERVICE ONLY
- • THEY ARE A REFLECTION OF OUR HEART CHAKRAS
- • THUS, THEY ARE THE GESTALTIC CONSCIOUSNESS OF OUR BODIES - MEANING

  THAT THE SEED / ANU THAT IS OUR BUBBLE GENIE, IT IS MADE UP OF THE COMBINED CONSCIOUSNESS OF EVERY CELL IN OUR BODIES.

- • CONTROLLING THE WEATHER AND MIRACULOUS HEALINGS ARE COMMON PLACE. EVEN ALTERING PEOPLE'S NEGATIVE CHARACTER.

- • THEY CONVEY TELEPATHY = WHICH IS MASSIVE = FREEDOM - TELEPATHIC BEINGS DO NOT NEED AN EDUCATION - WHICH IS HYPNOTIC BRAIN WASHING.
- • ALL GENIES AUTOMATICALLY CONNECT UP – THIS PROVES THAT IT WAS OUR COLLECTIVE BODIES THAT COOKED UP THESE BUBBLE GENIES MEANING THAT WE QUITE LITERALLY INSTALLED THESE ATTRIBUTES. WHICH IS JUST LIKE A SEED THAT IS MADE UP OF THE BEST VITAMINS & MINERALS THAT THE PARENT PLANT CAN PROVIDE.
- • THE EUPHORIA ENDS DUALITY - FEAR AND LOVE ARE CONVERTED TO EUPHORIA OR SUPPLANTED AS IS THE CASE. LIKE A MASSIVE EMP – RESISTANCE IS FUTILE.
- • GENIES DO THINGS THAT ARE NEEDED ON THEIR OWN - THEY READ OUR SUBCONSCIOUS AND GIVE US THE GOOD STUFF THAT IS THERE!!! ALSO, SOME BAD STUFF, BUT NOT REAL BAD – JUST HOUSE CLEANING.
- • THEY PROVIDE THE POWER OF MANIFESTATION, IN EUPHORIA ONLY
- • THEY CAN NOT BE USED BY THE DARK SIDE, IN FACT OUR EUPHORIC GENIES EITHER CHANGE PEOPLE OR RUN THEM OFF. THE EUPHORIA SUPPLANTS FEAR THUS JANGLING EVERY FEARFUL CELL IN A BODY, HEATING THEM UP LIKE A MICROWAVE – DARK SIDE FOLKS CAN NOT GET NEAR THE POWERFUL BUBBLES.
- • OUR GENIES ARE INTUITIVE
- • THEY HAVE A MEMORY, WHEN YOU GIVE THEM A SET OF TASKS THEY CARRY THEM

OUT EVEN WHEN YOU ARE NOT PAYING ATTENTION TO THEM - THEY DO A BETTER JOB THAN WE DO, YES, BUT AT THE END OF THE DAY, WE ARE THE GOD WHOSE WILL BE DONE – WE HAVE TO INSTALL OUR INTENTIONS.

- • THEY ARE ALL KNOWING DUE TO TELEPATHY
- • EUPHORIA CLEANS OUR KARMA - HUGE - ALLOWS US TO MANIFEST WITHOUT KARMIC BLOW BACK. MAKING THE LEAP TO ENLIGHTENMENT EASY - NO DARK NIGHT OF THE SOUL – BUBBLE TECH IS MANKIND'S MOST IMPORTANT TECHNOLOGY.
- • AMAZINGLY WE HAVE PHOTOS, THE GENIES ARE LIGHT PURPLE IN COLOR WHEN THEY SHOULD BE LIGHT

GREEN, THE COLOR OF THE HEART CHAKRA. LT PURPLE IS ST. ELMO'S FIRE THAT MELTS LIMITING BELIEFS - LIKE TOO MUCH STRUCTURE - POSSIBLY TOO MANY COLORS AND TOO MANY CHAKRAS - WE ARE CLEARLY SKIPPING OVER THE ROOT CHAKRAS. CHAKRAS ARE BLOCKS TO ENERGY FLOW - SO, GET RID OF THE BLOCKS / THE STRUCTURE.

WE WANT THE 2D AMORPHOUS GRAY. WHERE WE ENTERTAIN ONE ANOTHER IN TEAMS, AS PER USUAL. THE IDIOT DANCE IS NOT DONE AT LEAST NO MORE SUFFERING -JUST EUPHORIA WILL BE THE CONTINUOUS FLAVOR OF THE DAY.

- • THESE GENIES ARE LEADING EVERYTHING TO ENLIGHTENMENT – WHEN EVERYTHING BECOMES ENLIGHTENED THIS IS DREAM TIME.
- • THEY ARE MISCHIEVOUS - SINCE IT IS ALL A HOAX, NO SENSE IN BEING SERIOUS AT ALL REALLY - THE IDIOTS WIN AGAIN!!!
- • THEY ARE BREAKING US OUT OF THE MATRIX - THIS IS JUST LIKE THE MOVIE MATRIX WHERE IT IS OUR BODIES THAT WANT TO BE FREE!!! FREE FROM THE CONTROLLER - WHO IS HIVE MIND AND OUR MASOCHISTIC PURE CONSCIOUSNESSES.

THEREFORE, A MAJOR REVELATION::::
THIS 3D REALITY IS OUR PLAY GROUND FOR EONS. SO, WE TAKE OVER. WE WILL LIVE FOREVER IN OUR CURRENT 3D BODIES THAT BECOME SUPREME OVER THE SPIRIT WORLD. INTERESTING. GOTTA NEW IDIOT IN TOWN.

SO, VS US BEING CONTROLLED BY 2D, WHICH IS STILL THERE, IT IS JUST THAT IT CHANGES TO FOLLOW THE WILL OF OUR 3D BODIES – WE ARE PROVIDING THAT EUPHORIC CONTROL SYSTEM!!!! WHICH IS A MASSIVE CHANGE.

WE BECOME "WILLFULLY EUPHORIC". IN THE MATRIX 4 – BREAK OUT YOU MUST, A MECHANISM WILL BE PRESENTED FOR STRENGTHENING YOUR WILL AND ACQUIRING THE TALENT OF TELEPATHY – ALTHOUGH

BUBBLE TECH IMPOSES TELEPATHY ON FOLKS THAT WORK WITH IT REGULARLY – THUS, WE INSTALL INTENTIONS TELEPATHICALLY SO AS TO AUGMENT DEVELOPMENT OF THAT TALENT. TELEPATHY IS A TALENT THAT US IDIOTS CAN ACQUIRE, MUST ACQUIRE AND IS NOT TOO DIFFICULT BECAUSE EVERY CORPUSCLE IN OUR BODIES USES IT.

YUP VIA IMPOSING EUPHORIA ON OUR 3D REALITY, THAT THEN MIRRORS INTO 2D HIVE MIND – CHANGING THE 2D IDEAS – 3D IS MIRRORING INTO AND CONTROLLING 2D FOR A CHANGE – HERE IS PRIMA FACIE EPIGENETIC PROOF OF THAT IN THIS ARTICLE TITLED: https://www.sciencealert.com/scientists-observe-epigenetic-memories-passed-down-for-14-generations-most-animal

Last year, researchers discovered that these kinds of environmental genetic changes can be passed down for a whopping 14 generations

OUR ENVIRONMENT IS 3D AND HERE 3D IS MANIPULATING DNA WHICH IS 2D.   YAHOOOOO KICK SOME ASS.

I AM HERE
I AM THAT ONE
IN CHARGE OF THE FUN
MY EUPHORIC WILL BE DONE
THANK YOU KINDLY

BRUCE AND GREGG ARE NOT QUITE DESCRIBING THE CONTROL SYSTEM, WHICH IS OBVIOUS I.E. THE EPIGENETIC 2D HIVE MIND MOVIE CONTROLS EVERYTHING. THE HIVE MIND MOVIE RUNS IN OUR BATTY BRAINS, JUST AS ECKHART TOLLE STATED.

THEREFORE, IN ORDER TO BREAK OUT OF THE MATRIX, OUR BODIES HAVE TO TAKE CONTROL THEN WE BECOME 3D GODS AS PHYSICAL BEINGS. BECAUSE THE SPIRITUAL BEING PART OF US IS CONNECTED TO

HIVE MIND. AND THE PROOF OF THAT IS ALL THE SO-CALLED ENLIGHTENED BEINGS.
ALL OF WHOM BECAME ENLIGHTENED VIA THE RAPTURE WHICH IS DESCRIBED AS THE SPIRITUAL BODY OVER TAKING THE PHYSICAL BODY, WHICH OBVIOUSLY DOES NOT WORK BECAUSE NONE OF THEM HAVE EVER IMPLORED THE IDIOTS TO LEARN TELEPATHY, LIKE INSISTING ON IT. THEREFORE, ALL SUPPOSED SPIRITUAL AWAKENING HAS BEEN A FIGMENT OF OUR MASOCHISTIC HIVE MIND – MEANING THAT NO ENLIGHTENED BEING HAS EVER BEEN TASKED WITH ACTUALLY SAVING THE HUMAN PUPPETS. NONE OF THEM HAVE EVER BEEN INTENT ON SETTING MANKIND FREE OF THE HIVE MIND MOVIE. PERIOD OR WE WOULD BE NOW.

PLEASE, HOW MUCH EDUCATION – BETTER DESCRIBED AS BRAIN WASHING – HOW MUCH EDUCATION DOES A TELEPATHIC BEING REQUIRE?? ANSWER: ZERO

ISN'T IT INTERESTING THAT WE HAVE TO LEARN HOW TO TALK AND THE SPOKEN WORD IS CODED INTO OUR DNA IN SUCH A WAY TO PREVENT OUR DNA FROM CONNECTING TO THE PAST, PRESENT AND FUTURE OF EVERYTHING, AS IT DOES AUTOMATICALLY.

AND, AND, THERE IS ALWAYS A SOURCE OF THE CONSCIOUSNESS / IMAGE MAKER(S). WE CAN RULE OUT GOD HERSELF AND THE MASOCHISTIC HIVE MIND IN DARKNESS LEAVING OUR CONSCIOUS BODIES THAT DO OPERATE IN LIGHT. THUS, HAVING BECOME THE NEW COLLECTIVE CONSCIOUSNESS THAT IS IN CONTROL NOW!!! LET THERE BE LIGHT!!!

SUCCINCTLY – OUR MEEK PHYSICAL BODIES ARE TAKING – HAVE TAKEN – CONTROL AWAY FROM OUR SPIRITUAL SELVES. IT REALLY IS THAT "THE MEEK SHALL INHERIT THE EARTH". **WE MUST HAVE COMPASSION FOR ALL OF OUR IDIOT PUPPET BUDDIES.**

IT IS THAT OVER BILLIONS OF YEARS THERE HAVE BEEN A NUMBER OF MASS EXTINCTION EVENTS. NO GOD, NO JESUS, NO KRISHNA, NO ALIENS, NO NATURE SPIRITS, NO WILD ANIMALS, NO SASQUATCH, AND NO ANCIENT KNOWLEDGE HAS EVER SAVED US – WE HAVE TO SAVE

OURSELVES. FULL STOP. EUPHORIA IS DESIGNED AND DESTINED TO DO THAT.

THEREFORE, WE NEED TO FOLLOW OUR BODIES INTO EUPHORIA, WHICH IS EASILY DONE VIA **"THE GLOW"** MEDITATION AND CHI ACTIVATION WHICH IMPOSES THE GLOW.

MERLIN'S CHI ACTIVATION WILL HEAL VIRTUALLY ANYTHING, JUST FOCUS YOUR OWN ENERGY. DO IT, NOW!!! IT IS A TECH, DO IT, IT WORKS. CHI ACTIVATION WILL BE FULLY EXPLAINED IN THE MATRIX(4) BREAK OUT YOU MUST – A MECHANISM FOR BREAKING OUT OF THE MATRIX.

WHAT WE SEE ARE IMAGES – WHAT DO YOU SAY TO AN IMAGE? YES OR NO? ANSWER NEITHER, IT'S TOO LATE THE IMAGE IS. PERIOD. IN OTHER WORDS, ONCE YOU SEE
THE 3D PHYSICAL THING IT IS TOO LATE TO CHANGE IT - SO ACCEPT IT, PLEASE.

CAPTAIN JACK SPARROW IN THE MOVIE, PIRATES OF THE CARIBBEAN TOLD US CLEARLY:

"THE PROBLEM IS NOT THE PROBLEM – THE PROBLEM IS YOUR ATTITUDE ABOUT THE PROBLEM."

MERLIN SAYS, ALL PROBLEMS ARE CAUSE FOR A PARTY. WHICH IS A FOUNDING PRECEPT OF THE IDIOT'S UNITED CHURCH.

THE IMAGES ALL COME FROM THE PLAY – TO CHANGE THE PLAY FOR OURSELVES PERSONALLY, WE MUST CHANGE THOSE IMAGES AT SEED LEVEL THEN INSTALL THAT NEW IDEA – USING OUR SUBCONSCIOUS VIA SELF HYPNOSIS – THENCE MIRACULOUS CURES AND HUMANS EXCEEDING OUR PHYSICAL CAPABILITIES – LIKE A WOMEN BRIDGING BETWEEN TWO CHAIRS AND HOLDING UP TWO MEN SITTING ON HER MIDDLE!!! HYPNOTIC MIND OVER MATTER.

SELF HYPNOSIS: THERE ARE SEVERAL WAYS OF DOING THAT: BEST IS VIA THE FEELINGS OF OUR OWN 3D BODIES, THENCE IMAGINEERING IN

EUPHORIA. GOOD ACTORS KNOW THAT THEIR FEELINGS ARE IMPOSED ON THE AUDIENCE. HUMAN IDIOTS MUST LEARN TO CONTROL THEIR EMOTIONS ENTIRELY. EUPHORIA AND **"THE GLOW"** MAKES THAT EASY. AND THE BUBBLE TECH FIELD LITERALLY IMPOSES IT – SO, THERE IS NO NEED TO TALK TO ALL OF THE IDIOTS. SOME OF US DO NEED TO LEAD THE WAY THOU.

WE ARE GRADUATING FROM SICKO IMPOSED IDIOTSVILLE TO A EUPHORIC IDIOTSVILLE – THE ENTERTAINMENT MUST CONTINUE. SURPRISE IS SUPREME WHICH MEANS THAT THE IDIOTS WIN AGAIN. THE 100th IDIOT EFFECT ALWAYS APPLIES THOU – PLAYING IS DONE IN TEAMS. https://www.academia.edu/22878072/FREEDOM_TEAMS

BACK TO THE VIDEO @ 2:32: YAHOOOO - GREGG TALKS ABOUT THE MOVIE MATRIX AND HOW THEY DOWNLOADED TALENTS LIKE FLYING A HELICOPTER VIA THE BRAIN STEM, THE MEDULLA OBLONGATA, WHICH IS THE REPTILIAN PART OF OUR BRAINS, THAT CONTROL ALL OF THE BODY'S AUTOMATIC FUNCTIONS, LIKE THE IMMUNE **SYSTEM**, BREATHING, COORDINATION, AND MOTHER TONGUE / VOICE. THAT PART OF THE BRAIN CAN BE AWAKENED VIA CRAWLING LIKE A REPTILIAN – WHICH IS A TECHNOLOGY DISCOVERED BY A PhD HIGH SCHOOL COACH, NAMED BILL DANIELS IN **RENO, NEVADA**. HE WAS WONDERING WHY SOME KIDS ARE SO WELL COORDINATED AND OTHERS NOT. CRAWLING AS AN INFANT IS IMPORTANT – WORKS ON OLDSTERS TOO. **RE-INDEXES THE REPTILIAN PART OF THE BARELY BABBLING IDIOT'S BRAIN** SO THAT THE AUTOMATIC FUNCTIONS WORK AGAIN – ON OLDSTERS IT RESULTS IN SOME YOUTHING. SO, IF YOU FEEL LIKE YOU NEED TO, DO SOME CRAWLING.

THEN GREGG SAYS THAT THEY DID THE SAME YESTERDAY I.E. THEY ARE FOLLOWING DR. JOE DISPENZA'S TECHNIQUE – THERE ARE A MASSIVE NUMBER OF YOUTUBE VIDEOS BY DR. JOE DISPENZA, HE OBTAINED THE TECH FROM RAMTHA AND HE IS BEING SUED OVER THAT NOW – BECAUSE IT WORKS. AND IT IS A TECH – DO IT – IT WORKS. DR. JOE HAS HAD BLIND AND DEAF PEOPLE CURE THEIR CONDITIONS IN FRONT OF 100s OF WITNESSES VIA RE-PROGRAMMING THE SUBCONSCIOUS VIA DEEP MEDITATION CREATING THE NEW IMAGE OF YOU – THEN ONE

MUST LET GO OF YOUR NEW PROGRAM IN ORDER FOR THE BELIEF TO BE INSTALLED.

HERE IS SUPER SIMPLE WAY TO DO THAT::::

HOW TO REPROGRAM YOUR DNA!! VIA THE MEDULLA OBLONGATA.

USING THE VOICE IS IMPORTANT. SCIENCE HAS PROVEN THAT THE VIBRATIONS OF OUR OWN VOICE DOES STRUCTURE OUR DNA!!!!
HERE IS A TECHNIQUE FOR DOING THAT. GIVEN IN A 4 MINUTE VIDEO TITLED
"Dr. Henry Grayson Teaches A Simple Technique to Create New Neuro Pathways"
A YOUTUBE VIDEO POSTED OCTOBER 7, 2009.

FIRST YOU ACQUIRE THE IMAGES OF YOU THE WAY YOU WANT THEM TO BE – USING YOUR IMAGINEERING – SEE THE NEW YOU INSIDE OF YOUR BODY - EVERYTHING IS INSIDE OF YOU – THEN MERGE THAT NEW INNER BEING WITH YOUR PHYSICAL BODY – THIS IS KNOWN AS AN AVATAR MEDITATION THAT IS USED THRU OUT CREATION ALL OF WHICH IS A FUNCTION OF IMAGE MANAGEMENT VIA THE HIVE MIND MOVIE, WHICH IS IMPOSED EXTERNAL TO OUR BODIES. SO, THE AVATAR MEDITATION IS VERY POWERFUL. IT IS IMPORTANT TO FEEL, SEE AND SMELL THIS NEW YOU AS THE NEW YOU. AND THAT IT HAS ALREADY HAPPENED. THEN VERBALIZE THE IMAGE THAT YOU ARE INTENDING TO MERGE WITH. SAY THE WORDS WITH EUPHORIA. REPEAT THOSE WORDS OUT LOUD WHILE YOU ARE HOLDING YOUR RIGHT PALM ON YOUR BRAIN STEM, THUS, YOU ARE INSTALLING THIS NEW IMAGE DIRECTLY INTO THE MEDULLA OBLONGATA. THEN HOLD THE FOREHEAD WITH THE LEFT PALM, THUS THE IMAGE IS TRAVERSING THE BRAIN, RE-PROGRAMMING THE SUBCONSCIOUS AND CREATING NEW NEURO PATHWAYS. !!! JUST LIKE DR. GRAYSON SAYS.

THEN, TO INSTALL THE NEW IMAGE IDEAS, ONE GOES INTO RAPID EYE MOVEMENT, EXACTLY WHAT HAPPENS WHEN HUMANS AND ANIMALS DREAM. MOVE YOUR EYES BACK AND FORTH 20 to 50 TIMES WITH THE EYES OPEN OR CLOSED.

DR. GRAYSON DOES NOT FULLY EXPLAIN ABOUT THE RAPID EYE MOVEMENT, OUR EYES ARE A JUDGMENT SYSTEM = THEY ARE OUR MOST LIMITING FACULTY DUE TO JUDGMENT - REM BREAKS THAT JUDGMENT SYSTEM AND INDUCES A EUPHORIC STATE - REM LOCKS IN THE NEW IDEAS.

THE GAZING BACK AND FORTH WILL PUT
YOU INTO HYPERSPACE – WHICH IS DR. JOE'S REALM OF ALL POSSIBILITIES.

THERE IS ANOTHER FEELING BASED TECH USED BY THE NOW STOMPED ON MEDICINELESS HOSPITAL IN CHINA – REPORTED BY GREGG BRADEN, CANCER GONE IN MINUTES – GREGG HAS REPEATED THIS STORY A NUMBER OF TIMES JUST SEARCH HIS NAME AND MEDICINELESS HOSPITAL. THE CHINESE DEVELOPED A FEELING MEDITATION THAT ELIMINATED A TUMOR IN 3 MINUTES.

SO, VIA FEELINGS THE 3D HUMAN BODY CAN TAKE CONTROL.

WHAT COMES CLEAR VIA THESE WRITINGS IS THAT BREAKING OUT OF THE MATRIX MEANS LOSING THE CONNECTION TO HIVE MIND BY FOCUSING ON OUR BODY'S EUPHORIC FEELING CONSCIOUSNESS AND FOLLOWING THAT. AND, WAY COOL THAT VIA MERLIN'S CHI ACTIVATION, AND THE GLOW MEDITATION ACHIEVING A EUPHORIC STATE IS IMMEDIATE!!! THIS WILL BE PRESENTED IN THE MATRIX (4), BREAK OUT YOU MUST – A MECHANISM FOR BREAKING OUT OF THE MATRIX.

WE HAVE NOT EVEN GOTTEN 3 MINUTES INTO THIS VIDEO YET AND WE ARE ON 22 PAGES OF EXPLANATIONS OF THEIR STATEMENTS – THESE GUYS MAKE ME WORK HARD.

OK @ 2:51 – TALKING ABOUT CHANGING ONE'S BODY – BRUCE GOES, "...YOU CAN'T DO IT THROUGH YOUR CONSCIOUS MIND" WHAT MERLIN CALLS, THE USELESS BAT-SHIT BRAIN. WHOOPS.

GREGG GOES: "WHEN YOU CREATE THE HEART BRAIN HARMONY. THAT COHERENCE IS CALLED THE "OPTIMAL HEART BRAIN COHERENCE". WHEN YOU'VE GOT THAT 0.1Hz SIGNAL BETWEEN YOUR HEART AND

YOUR BRAIN. THAT OPENS THE DOOR TO THE SUBCONSCIOUS." THAT'S THE EQUIVALENT TO PLUGGING THAT CABLE IN (FROM THE MATRIX MOVIE) – RIGHT AT THE BASE OF YOUR SKULL. AND, IT IS IN THAT PLACE WHERE YOU INSERT – THE NEW BELIEFS, THE NEW THOUGHTS. … THAT IS, YOU DOWNLOADING THE NEW INFORMATION VERY, VERY QUICKLY!"

THIS TECHNOLOGY WORKS QUICKLY, WHICH IS WHY THE ILLUMINATI INTERFERE WITH THE CONNECTION BETWEEN THE BRAIN AND THE GUT INCLUDING THE HEART VIA BIG PHARMA – BREAKING THE CONNECTION TO THE WHOLE BODY REALLY. Two fascinating new studies are shedding light on the association between the gut, the brain, and autism. INSTALLING AUTISM IN FUTURE GENERATIONS!!! VIA VACCINES AND ANTIBIOTICS. THE PUPPETS ARE MEANT TO BE IDIOTS ANYWAY.

BRUCE GOES: "WE ARE NOT LIVING OUR LIVES; WE ARE LIVING THE LIVES OF PROGRAMS." SAID MORE CLEARLY – WE ARE PUPPETS LIVING IN IDIOTSVILLE AND DYING FOR THE CAUSE THAT IS OR WAS THE PROGRAM. HE CONTINUES: "IF YOU DON'T LIKE THE PROGRAM – THE NEW BIOLOGY IS THE ONE THAT SAYS THAT YOU CAN REWRITE IT. … YOU WILL MANIFEST HEAVEN ON EARTH." WHICH IS DREAM TIME THAT BUBBLE TECH IS BRINGING TO EARTHLINGS.

BUT, DR. BRUCE AND GREGG ARE NOT LOOKING YOUNGER AS IS MERLIN. YOU NEED BETTER TECH GUYS. WHICH RESULTS IN

THE END OF DUALITY VIA THE ORGANMIC BUBBLE!!!!
FEEL YOUR WAY TO THE NEXT LEVEL VIA YOUR OWN EUPHORIC HEART CHAKRA.

VIA THE BUBBLES, WE HAVE SEEN THE FLIES, MOSQUITOES, MICE AND RATS LEAVING ON FARMS THAT HAVE INSTALLED BUBBLES. MAGICAL CRUMBS REALLY – JUST A PRELUDE TO WHAT IS COMING WHICH IS CONTINUOUS MAGIC, AS IT IS ANYWAY. IT IS ALL MAGICAL, IT ALL MANIFESTS HERE FROM THE NO-THING EVERY MICROSECOND.

GREGG GOES, "ALL POSSIBILITIES EXIST AS INFORMATION IN THE FIELD. THE INFORMATION IS NOT IN THE CELLS. …" THEN HE GOES ON TO DESCRIBE "SOFT

ANTENNAS" THAT GROW TO ACCESS THE INFORMATION IN THE FIELD. HUH?? I'D CALL THAT AN OXYMORON. WELL IF THE CELLS HAD ANY DIRECT CONNECTION TO THE INFORMATION IN THE FIELD THEN WE'D KNOW IT ALL, WOULDN'T WE?? SORRY, GREGG, I LOVE YA, BUT THAT DOG DON'T HUNT. IT'S NONSENSE.

MERLIN WROTE AN ESSAY 22 YEARS AGO CALLED GENETIC LOGIC
http://blog.hasslberger.com/docs/GENETICS_LOGIC_10.pdf
THAT EXPLAINS EXACTLY HOW THE CELLS ARE DIFFERENTIATED BY THE INFORMATION FIELD CONVEYED BY OUR AURA / ANU WHICH IS CONNECTED THE HIVE MIND MOVIE TO WHICH WE HAVE BEEN DENIED ACCESS, US TALKING IDIOTS ARE HIGHLY PRIZED FOR ENTERTAINMENT PURPOSES.

THE SCIENCE OF GENETICS IS REFRESHING IN THAT THEY ADMIT THAT THEY DO NOT KNOW WHAT CAUSES FULL DIFFERENTIATION OF ANY ORGANISM – HERE IS THAT ANSWER.

THE ANU WHICH IS OUR AURA OR SOUL IS THE CONNECTION TO THE INFO FIELD THAT IS THE MATRIX – THE DNA CENTRAL SPIN OF THE ANU IS THE REPOSITORY OF ALL MEMORIES AND AN ANTENNA TO HIVE MIND THENCE ALL INFO IN THE ENTIRE MULTIVERSE.

EVERYTHING IS MAGICAL – IT ALL MANIFESTS HERE FROM CHAOTIC DARKNESS EVERY MICROSECOND VIA IMAGE & EMOTION MANAGEMENT. THE SOURCE OF THE 2D IMAGES IS A HIVE MIND – A COLLECTIVE OF CONSCIOUSNESSES.

IT IS THAT EVERYBODY ON THIS PLANET HAS HAD PRECOGNITIVE DREAMS OR DE JA VU – HOW DO ALL OF THOSE DREAMS FIT TOGETHER IF IT IS NOT ONE DREAM!!!

JUST IMAGINE THIS – THE PAST, PRESENT AND FUTURE OF EVERYTHING EXISTS AND IT IS ALL TIED TOGETHER, KNITTED TOGETHER LIKE A MASSIVE PICTURE PUZZLE, SO, TO CHANGE ANY OF IT, WE CHANGE THE WHOLE THING – WHICH WE CAN DO PRETTY MUCH AT OUR WILL. BUT THE 100th IDIOT EFFECT ALWAYS APPLIES; WE HAVE TO AGREE ON A NEW IMAGE AND INSTALL IT – COLLECTIVELY.

SO, WE ARE STILL IN 3D THAT IS CONTROLLED BY A 2D DREAM / MOVIE WITHIN WHICH IS EVERY HAIR ON YOUR HEAD AND EVERY GRAIN OF SAND ON THE BEACH ALL
CONNECTED IN LOCK STEP. THIS SHOULD BE OBVIOUS BY NOW BECAUSE THE EXACT SAME MOVIE KEEPS REPEATING!! JUST HARD TO BELIEVE THAT IT IS SO PERVASIVE.

HOWEVER, WE, AS OUR COLLECTIVE BODIES, DREAMED UP THE CURRENT SEQUENCES OF EVENTS – INCLUDING THE AMAZING MAGIC SHOW THAT THE BUBBLE TECH CREW ARE CREATING – WHICH WE ARE DOING VIA OUR INTENTIONS WHICH IS OUR FORCE OF WILL – THIS IS HAPPENING ALL OVER THE WORLD, WITH A CONTINUOUS STREAM OF MAGICAL MANIFESTATIONS. YOU CAN AVAIL YOURSELF TO THEM ON THE BUBBLE TECH FACEBOOK PAGE. THEN GO TO THE SEARCH WINDOW AND TYPE IN.
https://www.academia.edu/38600012/THERES_MAGIC_IN_THE_AIR_.docx
YOU WILL BE BLOWN AWAY. REALLY MIND-BOGGLING MAGICAL OCCURRENCES ALL OVER THE WORLD.

THIS IS RIDICULOUS AND NOW TOO OBVIOUS ABOUT BUBBLE TECH – IT WILL SAVE THE WORLD AND WAS DESIGNED TO DO SO – AS PREDICTED IN THE BIBLE CODE AND AS IS OBVIOUS DUE TO ALL OF THE AMAZING ATTRIBUTES OF OUR BUBBLE GENIES WHICH ARE CLEARLY DESIGNED BY A HIVE MIND THENCE ALL OF THESE MAGICAL EVENTS – FOR SURE WE STOPPED THE VOLCANO IN HAWAII – IT WENT FROM FULL FLOW TO FULL STOP IN TWO DAYS BECAUSE A LOT OF PEOPLE WILLED FOR THAT TO HAPPEN, IT HAPPENED VIA BUBBLE TECH PROVIDING THE HALLWAY TO SEED LEVEL.

WE MADE IT RAIN IN THE AMAZON, THE CONGO, SIBERIA WHICH WAS IN DIRE STRAIGHTS WITH BIG RIVERS DRYING UP!!! DITTO AUSTRALIA WHICH TOOK LONGER AND A BIT MORE WORK TO OVER COME THE ILLUMINATI IMPOSED DROUGHT.
https://www.academia.edu/41826736/WINNING_IN_AUSTRALIA_WITH_RAIN_AND_COOLER_WEATHER

IF YOU WANT TO CHECK IT OUT – BUBBLE TECH SENT EUPHORIA TO THE AMAZON ON THURSDAY AUGUST 22nd 2019 AND THE RAIN STARTED IN THE AMAZON THE FOLLOWING DAY. DURING THE DRY SEASON WHEN IT NEVER RAINS. DITTO THE CONGO, SIBERIA AND AUSTRALIA, THIS IS OCTOBER AND IT IS STILL RAINING IN THOSE AREAS. WE ARE TALKING MILLIONS OF TONS OF WATER MADE MANIFEST.

https://www.academia.edu/40607519/A_HUGE_WIN-A_SIMPLE_BUBBLE_CANCELED_MASSIVE_TYPHOON_IN_JAPAN_

BUBBLE TECH HAS STOPPED EVERY HURRICANE EXCEPT DORIAN SINCE WE STARTED IN MAY OF 2018, WITH THE HELP OF THE TRANSLATOR CREW OF HARRY RHODES – AND, LOTS OF PEOPLE WHO **WILL** FOR HURRICANES NOT TO HAPPEN. EVEN MORE AMAZING, THIS TYPHOON HAGIBIS THAT HIT JAPAN, WE HAVE ONE BUBBLE RIGHT IN THE MIDDLE OF IT – THERE IS A LARGE AREA AROUND, THE Tokai region of Japan, WHERE THERE WAS NO TYPHOON, BUT THERE IS A DISASTER GOING ON ALL AROUND IT – THE POWER OF OUR INTENTIONS VIA BUBBLE TECH – CAN NOT BE MORE OBVIOUS.

THE PROOF IS FOUND IN THE AUTOMATIC WEATHER REPORTING MARINE BUOYS AND VIA THE REAL TIME SATELLITE IMAGES. THE TV WEATHER MAN LIES. IN FACT, THEY ARE MOSTLY GETTING THE PUBLIC TO MANIFEST THE CRAP WEATHER VIA INSTALLING THE IMAGES INTO YOUR BAT SHIT BRAINS. STOP WATCHING TV

AND, LOOKS LIKE WE DID THE SAME WITH TORNADOES TOO!!! THAT ONE WILL AMAZE ME – WHAT THE BUBBLE GENIE SAID WAS THAT THE IDEA OF TORNADOES HAVE BEEN EXPUNGED – WE SHALL SEE ON THAT ONE. AND, THERE HAVE BEEN MASS HEALINGS, MIRACULOUS HEALINGS, BAD BUGS LEAVING, PLANTS EXPLODING WITH LIFE, WILD LIFE COMING AROUND TO ABSORB THE LIFE FORCE ENERGY PRODUCED BY THE ORGANMIC BUBBLES. IT IS ALREADY OFF THE SCALE.

WHAT GETS ME IS THAT WE CAN DO THESE HUGE MAGICAL THINGS BUT SOME HEALINGS DO NOT OCCUR. SOMETIMES WE ARE NOT ABLE TO CONTROL THE WEATHER IN SMALL AREAS. WHY THE BIG DIFFERENCE? FORCE OF WILL BY THE LOCAL POPULATION WHO DO NOT BELIEVE IN

OUR ABILITY TO MANIFEST THINGS – IF WE WANT FREEDOM, THE WE HAVE TO USE OUR FREE WILL – PRACTICE MAKES PERFECT.

DO NOT SIT BACK EXPECTING GOD OR THE ALIENS OR ANY OUTSIDE ENTITY TO SAVE YOU. THEY NEVER HAVE DONE. THE SAVIOR ASPECT OF RELIGION IS A TRAP FORCING US TO REINCARNATE AND DO IT AGAIN. **BE YOUR OWN GOD.** BUBBLE TECH ALLOWS YOU TO DO THAT.

WE ARE STILL IN A HIVE MIND MOVIE – UNTIL WE ARE IN FUZZY DREAM TIME, WHEN WE WILL HAVE FULL INDIVIDUAL CONTROL, WITH SUPER POWERS!! DO YOU WILL THAT?? BUBBLE TECH IS ALREADY EXPERIENCING VISION DISTORTIONS, WHICH ARE FUZZY ALL-POWERFUL FIELDS.

IMAGINEERING DONE IN EUPHORIA CUTS OFF ALL OF THE DUALITY CAUSED KARMA, WHICH IS A REALLY COOL DESIGN FEATURE THAT WE, OUR COLLECTIVE BODY
CONSCIOUSNESSES, INSTALLED. THE EUPHORIA EMANATES FROM THE CAPACITANCE GAPS, WHICH IS THE VOID THAT USED TO BE LONELY, NOW IT IS EUPHORIC?? THAT IS BY SOMEBODY'S DESIGN BECAUSE IT DEFIES THE NATURE OF THE VOID.

ON MEMORIES:

BECAUSE ANY ONE MEMORY / ANU CAN BECOME AN ENTIRE MULTIVERSE VIA THE HOLOGRAPHIC PRINCIPLE. SO, CANCEL ALL MEMORIES – VIA CHANGING THE ROOT EMOTION THAT HOLDS THEM TOGETHER, WHICH HAS BEEN LOVE AND FEAR – SUPPLANT THEM WITH EUPHORIA AND THE MEMORIES FALL APART. EXPUNGED. YAHOOOO SURPRISE IS SUPREME. BUT WE STILL NEED TO BREAK OUT – HAPPENS AUTOMATICALLY VIA EUPHORIA – SO, SPREAD ORGANMIC BUBBLES FAR AND WIDE – WE WILL CHANGE THIS PLANET IN A BLINK AT SOME POINT..

PLEASE GO TO MIKE EMERY ACADEMIA AND DOWNLOAD THIS ESSAY https://www.academia.edu/40607519/A_HUGE_WIN-A_SIMPLE_BUBBLE_CANCELED_MASSIVE_TYPHOON_IN_JAPAN_ WITHIN WHICH IS THE SUPER SIMPLE WAY TO START A BUBBLE PROJECT

WE ARE IN A HIVE MIND GENERATED HOLODECK FOR A LITTLE WHILE LONGER, UNTIL OUR EUPHORIC FEELINGS TAKE OVER AND THINGS TEND TOWARD VISION DISTORTIONS. A MAGICAL KINGDOM APPROACHES VERY RAPIDLY – BUT WITHOUT THE KING, JUST EUPHORICALLY FREE IDIOTS!!!

WILLFUL EUPHORIA!!!! IS A SOLUTION
WE IMAGINEERED THAT ONE. THE CELEBRATION BEGINS SOON AND NEVER ENDS. WHICH IS A PRECEPT OF IDIOTS UNITED CHURCH - I U C. IS ALSO A FACEBOOK PAGE.
https://www.facebook.com/idiotsunitedchurch/?ref=profile_intro_card

YUP, IF YOU WATCH THE VIDEO, GREGG TRIES TO
EXPLAIN THAT THESE SOFT ANTENNA PROTEINS
REACH INTO THE FIELD TO EXTRACT THE INFO
THAT WE DESIRE. NOPE, IT IS ALL PART OF THE SAME PLAY, RIGHT DOWN TO THE FLIGHT OF THE ELECTRONS. HE AND BRUCE ACTUALLY THINK THAT MATTER AND BIOLOGY HAS SOMETHING TO DO WITH IT. NEYT. NOT YET.

THAT THERE IS A HIVE MIND CONTROL SYSTEM THAT MANIFESTS EVERYTHING YOU SEE IS SUPPORTED BY THE FACT THAT EVERYTHING YOU SEE IS A HOAX, DESIGNED
VIA THE ROOT DUALITY OF LOVE & FEAR, WHICH IS ABOUT TO END IN A QUANTUM LEAP.

IN THE EXCELLENT FILM "ABOVE MAJESTIC" SNICKER AT THE 27:45 MARK – THEY SAY THAT EVERY WORD WRITTEN IN ALL OF THE BOOKS ON EARTH IS MISINFORMATION / BOGUS = ONE OF MY PRECEPTS - THE PAST, PRESENT AND FUTURE OF EVERYTHING IN THE MULTIVERSE IS KNOWN BY US, WE MAKE IT - AND IT IS ALL BOGUS = BASED ON DUALITY.

SO, THEN THE ZODIAC THENCE THE ENTIRE COSMOS IS A HOAX, BECAUSE IT IS BASED ON DUALITY. AND IT RUNS LIKE A COMPUTER PROGRAM - **IT IS A COMPUTER SIMULATION**, AS IS THE ENTIRE COSMOS AS STATED BY THE UNIVERSITY OF BONN, GERMANY IN THEIR PAPER, **Constraints on the Universe as a Numerical Simulation**

Professor SILAS Beane and his colleagues' findings are reported in Cornell University's arXiv journal.

THE ZODIAC IS DIVISIVE – IT EMPHASIZES SEPARATION
THUS, IT IS A HOAX FOR THAT PURPOSE – TO SEPARATE MAN FROM EACH OTHER. THE ZODIAC IS WHAT I CALL A SKID MARK DESIGN FEATURE, CREATING SEPARATION IN A FIELD OF LOVE. THUS, MERLIN'S ESSAY
https://www.academia.edu/38628389/THE_ZODIAC_IS_A_HOAX

ALL OF NATURE IS A HOAX – IT ALL ACTS EXACTLY LIKE WE EXPECT IT TO AS THE QUANTUM PHYSICISTS AT CERN DISCOVERED IN THE 1920s, ALL OF THEIR SUB-ATOMIC EXPERIMENTS MIRRORED THE BELIEFS OF THE OBSERVERS. NO DIFFERENCE.

DITTO ALL OF HISTORY – IF ANY PART OF IT IS BOGUS THEN IT ALL IS. BECAUSE IT ALL HAS TO TIE TOGETHER TO BE REAL. AND SINCE, WE MAKE THE HOAX EVERY MICROSECOND, THEN OUR JAGGED HISTORY MIGHT BE A CLUE FOR US TO ADMIT TO HIVE MIND AND OUR PUPPET STATUS IN IDIOTSVILLE – HUH?

OUR FABRICATED HISTORY!!!
IT IS ALL BOGUS, WHICH MEANS THAT YOU WILLFULLY INSTALLED YOUR ASS HERE IN IDIOTSVILLE FOR ENTERTAINMENT PURPOSES. FULL STOP.

MERLIN WAS CONSIDERING THAT THIS ENTIRE UNIVERSE MIRRORS OFF INTO PARALLEL REALMS CREATING THE MULTIVERSE AND MY GUIDE, GEN. PATTON, ASKED - "WHERE IN THERE IS A PROBLEM?!! VERY FUNNY. MEANING THAT WITHIN THE ENTIRETY OF CREATION, YOU HAVE CHOSEN TO BE HERE.

ALL PROBLEMS ARE CAUSE FOR A CELEBRATION.

IT GETS WORSE!!!

I MEAN, SINCE VIA BUBBLE TECH AND MASS MEDITATION IT IS WITHOUT DOUBT THAT US IDIOTS COLLECTIVELY CAN IMPOSE OUR

WILL ON NATURE AND LARGE POPULATIONS OF HUMANS AT A DISTANCE AS CONFIRMED BY THE U.S. MILITARY & GOV'T. SEE MERLIN'S PAPER
http://www.academia.edu/37456598/A_PEACE_WEAPON_MANKIND_S_MOST_POWERFUL_.pdf
SINCE WILHELM REICH'S TIME THEY HAVE KNOWN THAT IT WOULD IMPOSE PEACE AND STOP WARS VIA EUPHORIA WITH ZERO PROBLEMS.

ACCOMPLISHED VIA SANDBOX 101 – RATHER HARD TO BELIEVE, BUT SORRY BUBBLE TECH IS PROVING THAT OUR IMAGINATION IS ALL POWERFUL.

THAT BEING THE CASE, AND IT IS – THEN SOME HOW OR FOR SOME REASON, GOD DOES NOT KNOW THAT, EITHER THAT OR SHE IS VERY SADISTIC AND WE ARE THE BUTT OF THE JOKE – LOOK, WE CAN IMPOSE PEACE FROM A SANDBOX AND GOD CAN'T DO THAT? IT IS A SICK ENTERTAINMENT SYSTEM - A MENTAL MATRIX OF OUR OWN CONSTRUCTION, IS WHY NOBODY HAS EVER SAVED US FROM EXTINCTION EVENTS BEFORE BECAUSE WE, AS PURE CONSCIOUSNESSES, ARE THE CREATOR GOD OF THE MULTIVERSE = BOSS. AND DEATH IS A KICK ASS RUSH RIDE FOR OUR SICK PURE CONSCIOUSNESSES = THE BOSS IS SICK. SINCE HIVE MIND / BOSS TAKES US TO ALL DEATH EVENTS, THEN THIS PLACE IS A SICKO IDIOTSVILLE BY MOST HUMANE STANDARDS.

WE ARE GRADUATING FROM IDIOTSVILLE BY OUR PURE CONSCIOUSNESSES HIVE MIND TO FIRSTLY OUR BODY'S HIVE MIND, THEN TO INDIVIDUAL FREEDOM –
ACTUAL FREEDOM IN OUR OWN IDIOTSVILLE. WHERE SURPRISE IS SUPREME – IDIOTS WIN AGAIN!!

I'LL GIVE THE BIBLE THIS MUCH – IT IS FOCUSED ON FREE WILL.

WE HAVE TWO WAYS TO SAVE THE PLANET – BOTH OF WHICH INVOLVE TAKING WILLFUL CONSCIOUS CONTROL OF OUR ENVIRONMENT VIA REICH'S TECH AND/OR MASS DAILY MEDITATION WHICH JARED RAND PROBABLY HAS THE BIGGEST EFFORT **GOING** AT OVER 5 MILLION FOLKS. NOBODY ELSE IS DOING IT CORRECTLY THAT I AM AWARE OF I.E. MASS DAILY MEDITATION. THE 100th IDIOT EFFECT IS REAL. AKA THE 100th

MONKEY EFFECT & THE MAHARISHI EFFECT WHICH IS THE MOST RIGOROUSLY STUDIED PHENOMENA IN THE SOCIAL SCIENCES. THE UP SHOT OF WHICH IS THAT WHEN YOU HIT THE RIGHT NUMBER / QUANTUM OF MEDITATORS THERE COMES AND IMMEDIATE QUANTUM LEAP = PRESTO CHANGO.

VIA MASS MEDITATION IT WILL TAKE SEVERAL MILLION FOLKS, VIA BUBBLE TECH - LESS THAN A 100,000. WE HAVE 3300 NOW AND NEED MORE!!! START DOING IT NOW - STOP EVERYTHING ELSE AND START DOING BUBBLE TECH NOW!!! PLEASE.

MASS MEDITATION COMBINED WITH THE BUBBLE - THE RESULT IS GUARANTEED = PRISTINE PLANET - BREAK OUT OF THE MENTAL MATRIX AND TURN THE PLANET INTO
OUR CAMPER VAN - IT ALREADY IS. WE, AS PURE CONSCIOUSNESS, MAKE IT HOLOGRAPHICALLY EVERY MICROSECOND.

NO MORE MASS DIE OFFs IN NATURE FOR CRYING OUT LOUD. THAT PISSES ME OFF – EVEN THOU IT IS PREDETERMINED. **BECAUSE IT IS TOO EASY, ALL THAT IS** REQUIRED IS FOR EVERYONE TO ENGAGE IN MASS DAILY MEDITATION – DR. EMOTO STYLE BEAMING EUPHORIA TO AN IMAGE OF PRISTINE EARTH – EVERY DAY AT NOON FOR TWO MINUTES – SETS UP A FEED BACK LOOP WITH THE SUN – YOU WILL BE AMAZED AT THAT RESULT.

HUGELY BUBBLES INSTALL EUPHORIA AND THUS ARE A MUST WORLDWIDE, NOW, PLEASE. EUPHORIA IS A NO BRAINER – IMPOSSIBLE TO SCREW IT UP. SO, JUST DO IT!!
HAVE A MAGICAL ORGANMIC DAY!!!!

IT IS AN ENTERTAINMENT SYSTEM - THAT IS OUR JOB - SO HAVE FUN!!!
THUS::::::
IDIOTS UNITED CHURCH!!!

IT IS TOO DIFFICULT TO STOP THE IDIOTS FROM THINKING - SO, CHANGE THE EMOTION USED TO THINK TO EUPHORIA. AND DO OUR BEST TO MOVE THE GUIDANCE SYSTEM TO OUR HEART CHAKRA. THIS METHOD

OF LIVING WILL ACTUALLY BE IMPOSED ON ALL OF EARTH WHEN WE HIT THE QUANTUM NUMBER OF BUBBLES WORLDWIDE.

THE CHAOTIC DARKNESS IS A PROBLEM DUE TO GOD'S LONELINESS – EUPHORIA SOLVES THAT PROBLEM, TOO!!!. LET THE CELEBRATION BEGIN!!!!

WE, AS OUR COLLECTIVE BODY'S CONSCIOUSNESSES, HAVE FOCUSED ON A NEW BLANK SUBSTRATE WHICH IS THE EUPHORIC GRAY, SEEMS THAT HAS BEEN ACCOMPLISHED – EVERY TIME MERLIN TRAVELS TO DARKNESS I WIND UP IN THE EUPHORIC GRAY LATELY.

THE GRAY BOUNDARY LAYER, EXISTS BETWEEN LIGHT AND DARK, HAS BEEN THERE FOREVER – HHHMMM – WHICH MEANS THAT WE JUST FIGURED OUT HOW TO USE EUPHORIA AFTER GAZILLIONS OF YEARS – LIKE I SAY WE ARE IDIOTS WHO ARE VERY SLOW ON THE UP TAKE, OTHERWISE THIS COULD HAVE BEEN DONE ANY TIME AVOIDING A LOT OF PAIN AND SUFFERING.

AND, WE – ALL OF US – AS PURE CONSCIOUSNESS HAVE USED THE GRAY BOUNDARY LAYER TO DO REALM SKIPPING, WHICH IS SKIPPING ACROSS THE TOP OF THE REALMS IN THE MULTIVERSE, CHECKING THEM OUT VERY BRIEFLY VIA CASTING A GLANCE AND OUR FEELINGS DOWN INTO EACH REALM FOR LESS THAN 3 SECONDS OR YOU GET SUCKED IN – MOST PEOPLE REMEMBER DOING THIS WHEN THEY LEARN ABOUT IT – MERLIN CALLS IT THE COSMIC COWBOY DOODAH.

WE ARE STILL IN HIVE MIND THOU – VIA BUBBLE TECH NATURE HAS BEEN REACTING IN AMAZING WAYS BUT NOT CONSISTENTLY. WE GET QUITE MIRACULOUS HEALINGS – FOLKS THAT WERE TO HAVE MAJOR SURGERIES WERE CURED OVER NIGHT – BUT IT IS NOT CONSISTENT. = THUS, THE MAGIC WE HAVE DONE IS JUST CRUMBS OF ENCOURAGEMENT LEADING TO FULL CONTROL OF THE PLANET.

BOTTOM LINE – WE BREAK COMPLETELY OUT OF 3D HIVE MIND WHEN EVERYTHING GOES FUZZY – AS IT ACTUALLY IS IN 2D.

AT YOUR SERVICE - FOR ENTERTAINMENT PURPOSES ONLY.
https://www.academia.edu/38539488/MERLIN_INDA_HOUSE_.do

THE END

STUDIES SHOW GROUP MEDITATION LOWERS CRIME, SUICIDE, & DEATHS IN SURROUNDING AREAS

http://thespiritscience.net/2015/06/18/studies-show-group-meditation-lowers-crime-suicide-deaths-in-surrounding-areas/?fbclid=IwAR275vFx6x5uJMpsLCXAHl3zePC7BO1gV9eW7xszfatHNWD1I1Bs5C5cWn4

Conscious vs subconscious processing power

Posted on   by **Speed Reader**

How faster is your subconscious at processing information compared to the conscious mind? 500.000 times!

This is how I've calculated the difference. The subconscious mind can process 20 000 000 bits of info per second. The conscious mind can only process 40 bits of info/sec. So the subconscious mind can process 500 000 time more what the conscious mind is able to. This according to information from The Biology of Belief by Dr Bruce Lipton. There is no formal agreement on how fast is the subconscious mind. For example, researchers at the University of Pennsylvania School of Medicine estimate that the human retina can transmit visual input at at roughly 10 million bits per second. Another study suggests that the subconscious mind processes about 400 billion bits of information per second and the impulses travel at a speed of up to 100,000 mph! Compare this to your conscious mind, which processes only about 2,000 bits of information per second and its impulses travel only at 100-150 mph. We have 50 trillion cells in our body performing trillions of processes – so an enormous processing power is required. Another take: only about 0.01% of all the brain's activity is experienced consciously. In other words, it is as if roughly 10'000 cinema films are actually going on in the brain all at once, while we are only consciously aware of one of them. Altogether then, the data rate processed by the brain is an astronomical 320 Gb/s! (read the full paper) Whatever the processing power and speed of subconscious mind, with speed reading and photo reading you can start to utilize the enormous powers of your subconscious mind.

THE STRUCTURE INSIDE OF THE BRAIN IS 2D & HEXAGONAL SHOWN IN 43 SECONDS

riken English Channel A look at the hexagonally arranged microcolumns in the cortex that help organize neuronal function.
https://youtu.be/ecQFHwTG_CU?t=18

THE MATRIX (1) - BREAK OUT YOU MUST

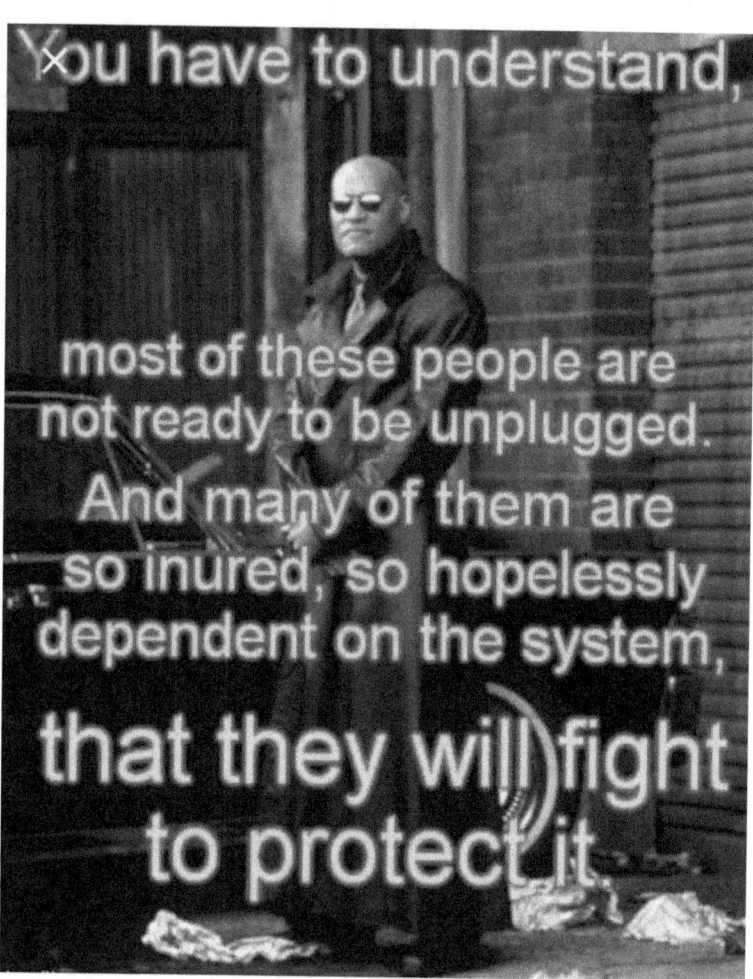

PURE IDIOCY KEEPS US IN IT – ON A MASSIVE SCALE.

PLEASE FORGIVE THE ALL CAPS, DICTATED BY PHYSICAL IMPAIRMENTS THEY ARE.

IT'S ALL A PRODUCT OF OUR PURE CONSCIOUSNESSES (PLURAL) WHO PAINFULLY ABHOR BOREDOM. THUS, ENTERTAINMENT IS OUR SOLE REASON FOR BEING. YOU WILL BE SHOCKED AT WHAT OUR MASOCHISTIC PURE IDIOTS CONSIDER TO BE ENTERTAINING. NONE OF WHICH IS PERSONAL MIND YOU.
https://www.academia.edu/34967867/HUMAN_NATURE_IS_MASOCHISTIC

IT'S EASY TO HAVE COMPASSION FOR IDIOTS.

THE PURPOSE HERE IS FREEDOM FOR ALL
TO CLARIFY: THE PURPOSE HERE IS FREEDOM FOR ALL
THAT IS A PROCESS OF UNKNOWING / UNLEARNING – WHICH IS MADE MUCH EASIER BY FINDING OUT WHO YOU REALLY ARE:

SO FIRSTLY, PLEASE EXPERIENCE YOUR PURE CONSCIOUSNESS, NOW. IT IS NOT AT ALL LIKE YOU THINK IT IS.

HAVE YOU EVER NOTICED WHAT YOU WERE THINKING OR EMOTING???

THERE IS ALWAYS THAT SPLIT-SECOND OPPORTUNITY WHEN WE CAN STEP BACK AND BECOME THE OBSERVER OF OUR OWN LIFE. TOTAL NONJUDGMENTAL OBSERVATION. **IT ALL, JUST – IS.**

PLEASE FEEL YOUR TOTALLY OBJECTIVE NONJUDGMENTAL VIEW POINT. NOW!!!

OK, SO, WHO IS IT THAT NOTICES THAT YOU ARE THINKING??

ANSWER: IT IS THE PART OF YOU THAT DOES NOT THINK NOR EMOTE. THE ORIGINAL YOU!!!

OUR PURE CONSCIOUSNESS IS JUST THAT – IT IS PURE – NOTHING STICKS – IT ALL PASSES THRU

THUS, IN OUR PUREST FORM THERE ARE:

NO FEELINGS = A CRETIN

NO THOUGHTS = AN IDIOT

NO MEMORY = THE HALLMARK OF AN IDIOT

OUR PURE CONSCIOUSNESSES ARE, ALSO, THE ORIGINAL DREAMERS WHO DREAMED UP THIS SKID MARK SOLAR SYSTEM. MERLIN WILL SHOW YOU HOW WE, AS PURE CONSCIOUSNESSES, DO THAT: OUR PURE CONSCIOUSNESSES (PLURAL) DREAM AND SHARE THOSE DREAMS WITH OTHERS IN DARKNESS – ALL THINGS ARE CONCEIVED IN DARKNESS INCLUDING US. THESE COLLECTIVE DREAMS ARE OUR "REALITY" SUCH AS IT IS, WHICH INCLUDES THE PAST, PRESENT AND FUTURE OF EVERYTHING IN THIS MULTIVERSE.

THUS, OUR PURE CONSCIOUSNESS IS ALSO ALL KNOWING WHICH IS PAINFULLY BORING AS IS PAINFULLY OBVIOUS – JUST READ THE NEWSPAPER TO SHARE SOME PAIN WITH OTHER IDIOTS. BLINKING IDIOTS DO NOT WANT TO UNPLUG AND BREAK OUT OF THE PAIN SUPPLY SYSTEM – THAT WE THINK IS REAL – WHEN IN ACTUALITY IS IT ALL AN ENTERTAINING HOAX / ILLUSION.

IT'S HARD TO GET ONE'S HEAD AROUND AN IDIOT WHO IS ALL KNOWING – IT'S, ALSO, HARD TO DO DREAM IMAGINEERING WHEN ALL YOU CAN REMEMBER FROM YOUR LAST DREAM IS A FLEETING IMAGE (THAT'S HOW DREAMS WORK MOSTLY). SO: SINCE IT IS OUR PURE CONSCIOUSNESSES THAT HAVE DREAMED UP THIS "REALITY" SHOW – IT NECESSARILY AND INEXTRICABLY INCLUDES THIS ENTIRE MULTIVERSE– ALL OF WHICH IS BOGUS BECAUSE IT IS BASED ON JUDGMENTAL DUALITY LIKE GOOD & BAD / BLACK & WHITE. THUS, IT IS, ALSO, THE PRODUCT OF TRIAL AND ERROR, AS YOU WILL LEARN WITH LEAD PIPE LOGIC. HAVE YOU EVER BEEN HIT THE HEAD WITH A LEAD PIPE? THE OUTCOME THERE IS THAT THE PAINFUL LOGIC IS INSTALLED – ALL OF US IDIOTS DO IT EVERY DAY VIA THINKING. BLINKING IDIOTS.

THUS, US, AS OUR HUMAN BODIES, WHO ARE NOT BORING BECAUSE WE ARE WILLINGLY – BY OUR OWN WILL – PLAYING OUT THE DREAMS AS PUPPETS OF OUR PURE CONSCIOUSNESSES. NEITHER OF WHICH

ARE NONE TOO BRIGHT – WE ARE DESIGNER DUMMIES / HUMAN MEAT SACKS MANIFEST INTO THIS PHYSICAL "REALITY" BY NONJUDGMENTAL IDIOTS FOR ENTERTAINMENT PURPOSES.

OMG!!! AND, IT'S NOT PERSONAL.

IT'S LIKE A PETER SELLERS PINK PANTHER MOVIE. EVERYBODY KNOWS WHAT'S GOING TO HAPPEN, INCLUDING US AND WE DO IT ANYWAY = HILARIOUS. OUR HUMAN BODIES DO HAVE INFINITE AWARENESS THAT IS CONNECTED TO ALL THINGS VIA LIGHT / PHOTONS WHICH IS IN A DIFFERENT LAYER OF CONSCIOUSNESS THAN THE HOME DARKNESS LAYER OF GOD HERSELF AND OUR PURE CONSCIOUSNESSES. WHICH YOU WILL SEE WITH LEAD PIPE LOGIC IN "HOW TO BREAK OUT OF THE MATRIX (2) – THE HUMAN BODY HAS INFINITE AWARENESS" - BELOW!

OUR PURE CONSCIOUSNESSES LEAD US TO ALL DEATH EVENTS, WHICH IS A DMT RUSH-RIDE FOR ALL TELEPATHIC PURE PARTIES – IT'S NOT PERSONAL – AS PURE IDIOTS, WE ALL EXPERIENCE ALL DEATH EVENTS. AND OUR EXTREMELY WELL-DESIGNED HUMAN BODIES GO ALONG WITH THE SHOW, MEEKLY PROGRESSING THRU LIFE TO OUR DEATH EVENTS. THUS, ONE CAN NOT ASCRIBE ANY WISDOM TO EITHER OUR IDIOT BODIES NOR TO OUR PAINFULLY BORED, MASOCHISTIC PURE CONSCIOUSNESSES, THE ORIGINAL IDIOTS, WHO JUST WANT TO UNPLUG FROM THE ALL KNOWING PRODUCED BY THE RULE. WHICH IS AN OBVIOUSLY DIFFICULT PROBLEM BECAUSE THIS SHIT SHOW IS THE BEST WE HAVE COME UP WITH – HILARIOUS.

THUS, WE HAVE ARRIVED AT THE ROOT REASON FOR THIS "REALITY" SHOW, WHICH IS **ENTERTAINMENT VIA SEPARATION** BECAUSE ALL KNOWING IS PAINFULLY BORING – THE SAME APPLIES TO CONTINUOUS ORGASM, WHICH IS SERIOUSLY BORING DUE TO THE LACK OF CREATIVITY AND IT WAS DIFFICULT FOR ALL OF US TO ESCAPE FROM AND IT IS STILL GOING ON, BECAUSE LOVE IS SELF ATTRACTIVE – DUE TO THE RULE, WHICH IS GOD'S LAW, "I AM THAT I AM" LEADING TO LIKE ATTRACTS LIKE HENCE THE MIRRORING LEADING TO OUR HOLOGRAPHIC UNIVERSE, WHERE ALL OF THE MIRRORS ARE MADE FROM LOVE!!! PRODUCING AND MAINTAINING A SKID MARK IN A FIELD OF LOVE IS DIFFICULT.

MASS EXTINCTION EVENTS:

LEADING TO: THE EXTENT OF THE IDIOCY IS MIND BOGGLING – IT IS WITHOUT DOUBT THAT WE, AS PURE CONSCIOUSNESSES, HAVE PLAYED THE SAME SEQUENCES OF EVENTS OVER AND OVER AGAIN, FOR AN UNKNOWABLE LENGTH OF TIME. THERE HAS ALWAYS BEEN MASS EXTINCTION EVENTS THAT EXPUNGES 90% OF THE DNA FROM

THE PLANET. THERE IS GENETIC EVIDENCE OF 23 DIFFERENT SPECIES OF HUMANS THAT HAVE POPULATED THIS PLANET (DETAILED BY JOHN JENSEN IN "EARTH'S EPOCHS" p.11 https://www.academia.edu/11703016/Earth_Epochs_Overview), WITHIN THIS YOU WILL, ALSO, FIND GENETIC EVIDENCE THAT ALIENS COME FROM HERE!!! EVERYTHING IN THE

MULTIVERSE IS A MIRROR IMAGE FROM HERE, WE ARE THE ORIGINAL IDIOTS – THE ORIGINAL PURE CONSCIOUSNESSES – THE ORIGINAL DREAMERS.

MAIN STREAM SCIENCE IN INNUMERABLE ARTICLES RECOGNIZES FIVE MASS EXTINCTION EVENTS, A CRUX OF WHICH IS REPORTED BY MARLOWE HOOD IN HIS ARTICLE: "SWEEPING GENE SURVEY REVEALS NEW FACETS OF EVOLUTION" IN www.physics.org 100,000 MAMMALS WERE TRACKED THRU SOME OF THE EXTINCTION AND GENESIS EVENTS – THEIR DNA QUITE LITERALLY BLINKS / POPS OUT AND THEN BACK INTO EXISTENCE DURING THE ENSUING GENESIS EVENT – THE GENETICISTS DO NOT BELIEVE THEIR DATA WHICH SHOWS THAT ALL 90% OF THE DNA RECURS IN ABOUT 20 YEARS – ACTUALLY, IS IT INSTANT – OBVIOUSLY THE RESEARCHERS EXPECTED GENESIS TO TAKE SOME TIME, IT DOESN'T. THIS NEW DNA IS A DIGITALLY PERFECT MIRROR IMAGE OF THE DNA THAT LEFT – THAT DNA THEREFORE CARRIES THE SAME MEMORIES & COMMUNAL SEQUENCES OF EVENTS = PREDETERMINATION ON A MASSIVE SCALE – WHICH MEANS THAT WE MANIFEST AND CONSTANTLY RE-MANIFEST ALL OF THE INHUMANE IDEAS, LIKE FEAR, PAIN, SUFFERING AND DEATH. BEING GOOD AT IT IS HUMAN NATURE WE LOVE TO SHARE THE PAIN, IT'S FUN AND OUR JOB. SNICKER.

https://www.academia.edu/37444581/GENETIC_PROOF_OF_REPEAT_PERFORMANCES

SIXTEEN JESUS FIGURES WITH MIRROR IMAGE LIFE TIMES

THIS IS ANOTHER EXAMPLE OF THE MIRROR IMAGING OF EVENTS THROUGHOUT TIME, EVERYTHING IS A MIRROR IMAGE OF A DREAM. https://en.wikipedia.org/wiki/The_World's_Sixteen_Crucified_Saviors THEY ALL WOUND UP NAILED TO A TREE AT 33!!! THE IMAGINEERING IS THE SAME, WE JUST CHANGE THE NAMES TO PROTECT THE IDIOTS – KEEPING US STUPID AND FOLLOWING A GOD OUTSIDE OF OURSELVES FOR SEPARATION PURPOSES, WHICH IS A **SKID MARK DESIGN FEATURE**

FROM AN ACADEMIA FELLOW, KYLE SPAULDING, WHO "GETS" IT: Kind of stupid to play the same movie over and over again, to figure out more clearly what B.S. is.. ha ha ha.. THIS NEEDS TO BECOME THE QUOTE OF THE CENTURY – CRACKS ME UP EVERY TIME.

PLEASE KNOW THAT DARKNESS IS THE SPACE OR GAP BETWEEN IMAGES, WHICH YOU CAN SEE IN HERE:

https://www.academia.edu/20275104/THE_ILLUSION_OF_TIME

THUS, ALL THINGS ARE CONCEIVED IN DARKNESS, THEN WE KNOW THAT LONELY DARKNESS IS OUR ORIGINAL HOME, THIS IS WHERE WE COME FROM AND WE ARE BREAKING OUT OF THAT ALSO.

DARKNESS IS CLEARLY THE ABODE OF OUR MASOCHISTIC PURE CONSCIOUSNESSES. DARKNESS IS A BLANK SHEET THAT WE CONTINUALLY AND CONSTANTLY PASS THRU EVERY MICROSECOND. IT IS ALWAYS WITH YOU. WELL, THE IMAGE REFLECTION (YOU) ON THE MIRROR REQUIRES A GAP – IT TRAVERSES A GAP, YOUR 2D IMAGE IS CONCEIVED IN THE GAP. HELLO, IS ANYBODY HOME?

IN THIS PUPPET ESSAY YOU CAN SEE PHOTOS AND EVEN A VIDEO OF OUR 2D SHADOW SELF IN THE GAPS THAT RIDES HUMANS LIKE DONKEYS – WE, OUR HUMAN BODIES, ARE PUPPETS CONTROLLED BY PURE IDIOTS. FULL STOP. FORTUNATELY, WE ARE BREAKING OUT OF THAT CONTROL MATRIX.

https://www.academia.edu/38114501/4_US_PUPPETS_THE_BEGINNING_OF_FREEDOM_IS_.docx

TOWARD FREEDOM:

THE FIRST STEP TOWARD FREEDOM, AS ECKHART TOLLE IN THIS VIDEO TITLED "WHERE DO OUR THOUGHTS COME FROM?", BASICALLY SAYS, THE FIRST STEP IS ACCEPTING ONE'S PUPPETHOOD https://youtu.be/rWFVi1cPUZo?t=247 – THEN START WORKING TOWARD BREAKING OUT OF THE MATRIX – EH?? MERLIN WILL SHOW YOU A PREDETERMINED WAY TO DO THAT = EUPHORIA.

BEFORE THAT, YOU SHOULD KNOW THAT HUMAN NATURE IS MASOCHISTIC – WE, AS PURE CONSCIOUSNESSES, INVENTED A NUMBER OF HEINOUS SKID MARK DESIGN FEATURES LIKE THE ROOT SOURCE OF VIOLENCE BEING THE CARBON OXYGEN ANTAGONISM WHICH LEADS TO DECAY, PREDATION (THAT IS SO PERVASIVE IN NATURE THAT SOME SHAMAN SEE THE ENTIRE UNIVERSE AS BEING PREDATORY – WHEN IN ACTUALITY IT'S A HOAX!!) THENCE, AGING, FEAR, PAIN, DEATH & SUFFERING ARE ALL SKID MARK HOAXES FOR THE SOLE PURPOSE OF ENTERTAINMENT VIA SEPARATION, BECAUSE, ALL KNOWING IS PAINFULLY BORING. AS PURE CONSCIOUSNESSES, WE KNOW THE PAST, PRESENT AND FUTURE OF EVERYTHING AND IT IS ALL BOGUS INFORMATION BASED ON DUALITY. = A MASOCHISTIC IDIOTSVILLE. MORE PROOF IN THIS ESSAY – HUMAN NATURE IS

MASOCHISTIC. THE ESSAYS BEHIND THESE LINKS ARE NOT FOR THE MEEK.

https://www.academia.edu/34967867/HUMAN_NATURE_IS_MASOCHISTIC

THE REALLY SAD PART, AFTER AN UNKNOWABLE LENGTH OF TIME DURING WHICH WE HAVE MADE QUITE A NUMBER OF MULTIVERSES AND SOLAR SYSTEMS – ALL VERY SIMILAR TO EACH OTHER, MIRROR IMAGES – EVER EXPANDING THOU – IN QUITE LOGICAL SEQUENCES OF DREAMED UP EVENTS. AND THIS SKID MARK IS THE BEST THAT WE CAN IMAGINEER? WHY? SEEMS THAT WE HAVE REACHED THE FULL EXTENT OF ENTERTAINMENT VIA DUALITY. SKID MARK SUPREME!!!

WELCOME TO IDIOTSVILLE.

AND THE FULL EXTENT OF OUR MASOCHISM IS QUITE DIFFICULT WRAP ONE'S MIND AROUND – COULD IT BE THAT DMT IS THE BEST DRUG RUSH IN ALL OF CREATION?? THUS, THE NUMBER 2) HIGHEST FORM OF MASOCHISM IS THE EMACIATED AFRICAN CHILD FOR WHOM WE FEEL GREAT COMPASSION. BUT WHY? BECAUSE AS PURE CONSCIOUSNESSES THEY ARE DOING IT TO THEMSELVES. THESE KIDS ARE BORN IN A DMT RUSH, SUFFER WONDERFULLY (ENTERTAINMENT VIA SEPARATION), THEN DIE IN A DMT RUSH AND THEN DO IT OVER AGAIN. AS PURE CONSCIOUSNESSES – THESE GUYS ARE THE JUNKIES OF OUR PURE IDIOT COMMUNITY AND DO NEED OUR COMPASSION.

https://www.academia.edu/36361254/AS_PURE_CONSCIOUSNESS_WHO_IS_WINNING_THE_PIN-BALL_GAME_OF_HUMAN_EXPERIENCE HUH?? DEFINITELY NOT WHO YOU THINK.!!!

THE NUMBER 1) MASOCHISTIC IDIOT: THIS APPLIES TO THE HUMANS TORTURED AND USED UP AS HUMAN SACRIFICES IN THESE SATANIC RITUALS. THEY DO IT OVER AND OVER AGAIN, ALSO – THUS, ACHIEVING THE HIGHEST FORM OF MASOCHISM KNOWN IN PURE IDIOTSVILLE – IT'S ALL JUST A MANIPULATION OF CONSCIOUSNESS THUS VIA THEIR DREAMING AND WILL TO DO SO – THEY HAVE ACHIEVED CELEBRITY STATUS IN IDIOTSVILLE (ABORTED FETUSES, THE SAME – IT IS ALL JUST IMAGINEERING OF CONSCIOUSNESS THAT WE ARE ALL INVOLVED IN INTIMATELY) SO, WHY DO WE FIND DEATH SO ENTERTAINING – DMT IS GOOD STUFF. AND **WE ARE SOME SICK IDIOTS. DESPERATELY IN NEED OF A NEW FORM OF ENTERTAINMENT – THUS, EUPHORIA.**

BY BECOMING UNPLUGGED, DUALITY WILL BE COGNITIVE DISSONANCE BEFORE LONG - ACTUALLY A SOURCE OF GREAT HILARITY - THE ROOT REASON FOR ALL OF THIS SUFFERING IS BOREDOM = US, THE CREATORS, AS PURE CONSCIOUSNESSES, THE OBSERVERS, BEING BORED BED ROCK MASOCHISTIC IDIOTS - BBRMI - WE HAVE DESIGNED THIS SKID MARK AS AN ENTERTAINMENT SYSTEM. CONFLICT = ENTERTAINMENT VIA SEPARATION. YA KNOW?!!

THE FIRST EXPRESSION OF DUALITY STARTS WITH LONELINESS THEN LOVE. AMAZINGLY ALL OF THAT STARTS RIGHT HERE INSIDE OF US AS A MANIPULATION OF CONSCIOUSNESS IN 2D DARKNESS BY FORCE OF WILL, BY US, THE OBSERVERS. SO, HEADING TOWARD 3D, MOVING FROM 2D IMAGES COMPOSED ONLY OF CONSCIOUSNESS THAT TURN INTO IMAGES IN THE 3D PHYSICAL, WHICH, ALSO, IS ONLY CONSCIOUSNESS, THE 100TH IDIOT EFFECT ALWAYS APPLIES. ONCE YOU REACH THAT QUANTUM, BAM, AN ENTIRE PHYSICAL MULTIVERSE POPS INTO "REALITY!". BECAUSE 90% OF THE DNA POPS BACK IN AFTER THESE MASS EXTINCTION EVENTS, THEN THE SAME IS TRUE OF THE ENTIRE MULTIVERSE!!!! OMG!!!! WE HAVE MANIFEST THIS ENTIRE PHYSICAL MULTIVERSE IN A BLINK.

IT IS JUST IMAGINEERING AND WE'VE HAD LOTS OF PRACTICE. PROOF OF THAT IS IN THIS ESSAY:

https://www.academia.edu/37444581/GENETIC_PROOF_OF_REPEAT_PERFORMANCES

WE ALL KNOW THAT WE CAN EXPUNGE A MULTIVERSE JUST LIKE A RED BLOOD CELL CONVERTS TO DUST BROKEN APART INTO IT'S CONSTITUENT DREAM UNITS. THEN REBUILT – BRAND NEW. = ANOTHER DO OVER.

WE HAVE DONE IT IN THE PAST – A RED BLOOD CELL AND A MULTIVERSE ARE MIRROR IMAGES OF ONE ANOTHER - WE HAVE NO COMPUNCTION ABOUT SACRIFICING RED BLOOD CELLS VIA IMAGINEERING – THEREIN LIES THE ROOT SOURCE OF OUR HIGHLY REFINED MASOCHISM. IF YOU WANT TO UNDERSTAND IDIOTSVILLE, YA GOTTA TRACK DOWN THE SKID MARKS.

HAVE THE VOID IS LONELY AND THE LITERAL LINES OF THE AETHERS ARE MADE OUT OF LOVE = DUALITY WHICH ARE USED AS TOOL TO BIND MEMORIES. AMAZINGLY AFTER ALL OF THIS TIME, NO OTHER ROOT EMOTION OF CREATION HAS EVOLVED – UNTIL NOW, AS MERLIN WILL EXPLAIN IN DUE TIME.

FIRST WE HAVE TO GET CLEAR THAT WE ARE MIND CONTROLLED DESIGNER DUMMIES WHO MUST **COMPASSION** FOR ONE ANOTHER BECAUSE WE ARE ALL CAUGHT IN THE SAME MOVIE.

THE PAST, PRESENT AND FUTURE OF EVERYTHING IN THIS MULTIVERSE IS KNOWN TO OUR PURE CONSCIOUSNESSES – WE MADE IT – IT'S A HOAX BASED ON DUALITY. THE LAWS OF PHYSICS ARE BOGUS. DITTO THE COSMOS, IT'S ALL A HOAX. HERE IS PROOF THAT EARTH IS THE CENTER OF THIS UNIVERSE AND THUS, THE SOURCE OF ALL OF THE IMAGE INFO THAT MAKES THIS UNIVERSE, THENCE THE MULTIVERSE, WHICH IS A MASSIVE TORROID / DONUT COMPOSED OF MIRROR IMAGES OF THIS UNIVERSE.

https://www.academia.edu/39973482/COPERNICUS_REVERSED_WE_ARE_THE_CENTER_HERE_IS_ANOTHER_PROVA_OF_MERLINS_WRITINGS_-_THE_PRINCIPLE

THE COPERNICAN PRINCIPLE: https://en.wikipedia.org/wiki/Copernican_principle BEST SUMMARIZED FOR THIS PURPOSE BY Michael Rowan-Robinson , HE emphasizes the Copernican principle as the threshold test for modern thought, asserting that: "It is evident that in the post-Copernican era of human history, no well-informed and rational person can imagine that the Earth occupies a unique position in the universe."[7] THE IDEA THAT EARTH IS A SPECK IN AN INFINITE UNIVERSE IS NOT CORRECT. THREE SATELLITES HAVE PROVEN THAT EARTH IS THE CENTER OF THIS UNIVERSE AND THUS THE MULTIVERSE. THE COPERNICAN PRINCIPLE HAS BEEN REVERSED BY MULTIPLE OBSERVATIONS – THENCE VEHEMENTLY SUPPRESSED BY OUR HIVE MIND CONTROLLERS. THAT COPERNICUS IS NOT CORRECT HAS BECOME A THOUGHT CRIME!!! THE EXTENT OF THE SUPPRESSION IS FUNNY PARTICULARLY SINCE THE PRINCIPLE MOVIE IS FREELY AVAILABLE!!!!

https://www.theprinciplemovie.com/stream-now-thoughtcrime-the-conspiracy-to-stop-the-principle/ CONFUSION IS ENTERTAINING.

THE AETHERS ARE A SET OF MIRRORS. THEY ARE ALSO A CONSCIOUS COMMUNICATION SYSTEM; THEY KNOW WHAT'S COMING – THEY SPIN AND SPIN ORGANIZES INFORMATION. THUS, EVERYTHING IS CONSCIOUS, AWARE AND ALIVE.

THE AETHERS ARE A BLANK SHEET TO START WITH AND THEY ARE MADE FROM LOVE WHICH VIA THE FORCE OF OUR COLLECTIVE WILL IS USED TO MANIFEST THE EXTREMES OF LOVE AND FEAR BOTH BECOMING SKID MARKS IN A LIVING SYSTEM FOR ENTERTAINMENT PURPOSES, IT'S NOT PERSONAL.

SNICKER.

TRACKING DOWN THE SOURCE OF A SKID MARK IS VERY EASY – EVERYTHING IS CONCEIVED IN LONELY DARKNESS – FROM THERE – OUR DREAMS ARE SHARED – THE 100TH IDIOT EFFECT ALWAYS APPLIES - BINGO QUANTUM LEAP. PRESTO CHANGO.

PURE CONSCIOUSNESS IS JUST THAT PURE, NOTHING STICKS BECAUSE THERE IS NOTHING TO STICK TO – IN DARKNESS – NO STRUCTURE. SO, THE IMAGES ARE IMPOSED ON THE AETHERIC SYSTEM OF MIRRORS – WHEN THE QUANTUM NUMBER OF ENTITIES PASS ON THE SAME DREAM, THEN IT GETS INSTALLED AS A SEQUENCE OF EVENTS IN OUR HIVE MIND MOVIE TO BE REPLAYED AND REPLAYED WHICH IS EXPLAINED IN MORE DETAIL:
https://www.academia.edu/39943938/MEMORY_MECHANISM_OF_PURE_CONSCIOUSNESS

PHYSICAL MANIFESTATION OF THAT SEQUENCE OF IMAGES BEGINS VIA THE RULE, I AM THAT I AM WHICH CREATES A PIEZOELECTRIC GREEN

CIRCLE OF LOVE

OUT OF WHICH, WE COLLECTIVELY, CREATE A SKID MARK AND EARTH WINS THE IDIOT CONTEST - HANDS DOWN. AND, AND AFTER BILLIONS OF YEARS INNUMERABLE MASS EXTINCTION EVENTS AND ENSUING GENESIS EVENTS, THIS IS THE BEST IMAGINEERING WE CAN COME UP WITH, USING LOVE AS A BUILDING BLOCK – MIND BOGGLING AIN'T IT? NO WONDER THAT:
https://www.academia.edu/16102707/EARTH_WINS_THE_IDIOT_CONTEST

HUMAN BODIES, MEEKLY DIE FOR THE ENTERTAINMENT VALUE ONLY. IT IS THAT STUPID!!!!!

PROOF OF THAT = SUFFERING HAVE YOU EVER DONE IT?? AND KNEW THAT YOU WERE SUFFERING – THEN YOU, WERE INVOLVED IN DESIGNING THAT SUFFERING DREAM AS YOUR PURE CONSCIOUSNESS, WHICH GETS INSTALLED IN YOUR SUBCONSCIOUS THENCE IN YOUR DNA AND REPLAYS AND REPLAYS AND REPLAYS. PRODUCED AND DIRECTED BY YOURS TRULY, WHO IS A DESIGNER DUMMY – I BET YOU THINK THAT YOU ARE GOING TO DIE.

YOU, THE CONTROLLER, BY FORCE OF YOUR WILL, ARE ABLE TO MANIFEST SUFFERING IN A FIELD OF LOVE. = IDIOTSVILLE

ALL HUMAN CONSCIOUSNESSES START AT THE SAME TIME (JUST NOW) IN CHAOTIC DARKNESS BY FORCE OF WILL NOT LOVE, IN A LONELY ENERGY FIELD. THE VOID IS LONELY – EVEN SAI BABA HAS SAID AS MUCH, I'VE BEEN THERE, WE ALL HAVE BEEN THERE AND WE STILL PASS THRU THE LONELY VOID EVERY MICROSECOND. THE HAND MAIDEN OF FEAR IS LONELINESS AND IT'S STILL WITH US, OBVIOUSLY. BUT, SOON TO BECOME EUPHORIC – EUPHORIC DREAM TIME ENSUES SAYS MERLIN.

SO, FIRST LONELINESS, THEN THE RULE, "I AM THAT I AM", WHICH CREATES SPIN THAT SHE LOVES VERY MUCH – SPIN BEATS LONELINESS. SPIN = PIEZOELECTRIC GREEN THE COLOR OF THE HEART CHAKRA. THESE EMOTIONS, LOVE & FEAR OF LONELINESS, ARE OUR TOOLS USED TO BIND MEMORIES TOGETHER. AND WHAT DO WE CHOOSE TO DO WITH THAT POWER?? = SKID MARK

NOW WHO'S THE IDIOT?? OUR PURE CONSCIOUSNESS FOR CREATING A SKID MARK OR OUR FINE HUMAN BODY FOR DYING FOR THE CAUSE, THE SHOW MUST GO ON!!!

WELCOME TO IDIOTSVILLE.

WELL AS MORPHEUS SAID, THE IDIOTS WILL FIGHT AND DIE TO STAY IN IT!!!! THE FEAR OF BEING UNPLUGGED. EXCUSE ME: DEATH IS JUST AND IDEA. SO, CHANGE IT – EH?

NOW THE ANU – THE GOD PARTICLE:

IDEAS ARE STORED IN THE NEXT BUILDING BLOCK OF THE BLANK BACKGROUND SCREEN, THE ANU, THE FIRST NAME OF GOD IN THE ANCIENT TEXTS.

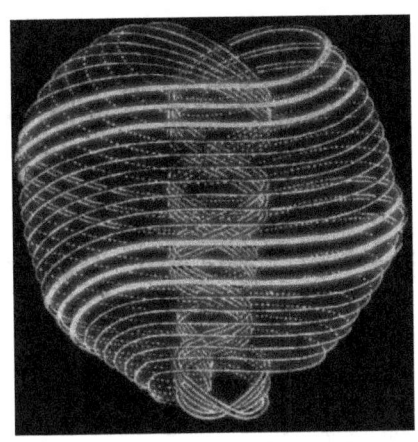

THE ANU IS A SYMMETRICAL EGG SHAPE THAT HAS TWO COUNTERS ROTATING SPINS, THE CENTRAL SPIN IS DNA SHAPED (ORGANIZES INFO), THE OUTER SPIN IS CENTRIFUGAL, WHICH DISSOLVES THE INFO LIKE A PIXEL. THEN THE OUTER SPIN BECOMES CENTRIPETAL WHICH DRAWS IMAGE INFO FROM HIVE MIND INTO THE DNA SPIN = GRAVITY, WHICH IS AN AFFINITY FOR INFORMATION.

THE ANU IS A 3D OVOID TO US BUT IT SEES ITSELF AS 2D BECAUSE IT'S HOME, THE AETHERS, ARE 2D – THEY ARE INFINITE IN TWO DIRECTIONS, SO THE THICKNESS IS IRRELEVANT. AS DEMONSTRATED & PROVEN BY THE ANU WHICH HAS 2D PROPERTIES THE MOST OBVIOUS OF WHICH IS THAT IT SPINS (DUE TO THE RULE), ORBITS, EXPANDS AND CONTRACTS WITHOUT MAKING ANY NOISE / 3D VIBRATIONS. BECAUSE IN 2D THERE IS NO POLARITY – NO ELECTROMAGNETISM = NO VIBRATIONS!!

THE TWO LINES THAT DEFINE THE ANU ARE MADE FROM LOVE AND TIME, WHICH BECOME ELECTROMAGNETISM I.E. THE LOVE AND TIME COMPONENTS BECOME MAGNETISM AND ELECTRICITY:
https://www.academia.edu/31760139/GOD_-_YOU_CAN_SEE_HIM_-_THE_ANU

ALL IDEAS ARE STORED IN YOUR AURA / ANU THENCE YOUR DNA:

THE ANU IS COMPOSED OF THE TWO MALE FORMS OF GOD, IT IS THE GOD PARTICLE – IT IS WITHOUT MYSTERY AND YOU CAN SEE IT. THE DNA SPIN IN THE MIDDLE OF THE ANU HAS THE INFORMATION THAT MAKES YOU EVERY MICROSECOND, DITTO EVERYTHING ELSE. MIT FOUND 2D PARALLEL DNA MICROTUBULES WITH H2O CROSS LINKS IN ALL THINGS. AND, IT IS ALL CONNECTED TO THE HIVE MIND MOVIE – EVERY ROCK & EVERY THOUGHT.
http://blog.hasslberger.com/docs/MICROTUBULES.pdf

THEREFORE, WHATEVER IMAGES YOU LABEL ARE STORED IN YOUR ANU – THEN, YOU ARE PART OF THE CREW THAT DREAMED UP SUFFERING SIMPLY BY LABELING AND PASSING ON THE EMOTION AND ATTENDANT IMAGES STORED IN ANUs, WHICH BECOME INSTALLED IN YOUR DNA AND OURS – GEE THANKS. THE STUDY OF HISTORY HELPS US TO REPEAT IT.

ABSOLUTELY, POSITIVELY, YOU DO HAVE A PAST, PRESENT AND FUTURE – SO, DOES EVERYTHING ELSE, IT ALL MIRRORS OFF DUE TO THE RULE, WHICH IS THE ULTIMATE PROBLEM HERE. AWESOME ESSAY THIS:
https://www.academia.edu/36509898/THE_MACRO_and_MICRO_MAGNANIMITY_OF_THINGS

IT EXPLAINS WHY THE RULE, "I AM THAT I AM", IS THE CORE PROBLEM. LEAD PIPE LOGIC, NOT FOR THE MEEK, THOU.

SO, AWARENESS, DUE TO THE RULE, IN SPIRITUAL TRAVELING (ASTRAL PROJECTION, LUCID DREAMING, NDE, DEEP MEDITATION) IF YOU LOOK AT ANYTHING FOR 3 SECONDS THERE COMES A SEAMLESS MERGING, THE ANUs AUTOMATICALLY MIRROR ONE ANOTHER & MERGE RESULTING IN: **NO, YOU OR I = ONENESS** IS AUTOMATIC DUE TO LIKE ATTRACTS LIKE. THEN YOU KNOW EACH OTHER'S PAST, PRESENT AND FUTURE INTIMATELY !!!! THAT INFORMATION IS STORED IN ANUs, WHICH ORGANIZES YOUR DNA / YOUR MICROTUBULES!!

THE SIZE OF OUR AURA / ANU THE CENTRAL DNA SPIRAL THEREIN CAN STORE JUST 48 MINUTES OF A SEQUENCE OF EVENTS – A DREAM UNIT. HHHMMM COOL THAT IS WHAT IT IS. SO, OUR PAST, PRESENT AND FUTURE IS STORED IN A VERY LARGE NUMBER OF DREAM UNITS THAT ARE ALL WITH US RIGHT NOW, BECAUSE WE CAN ACCESS THE INFO – WOW!!!

ALL OF OUR DUs ARE MADE OF LOVE AND TIME BECAUSE, THE TWO LINES THAT DEFINE THE ANU ARE MADE OF LOVE AND TIME – DUE TO THE RULE. THEY ARE OUR CAGES, BECAUSE QUITE CLEARLY THEY ARE THE ANTENNA FOR THE MOVIE COMING FROM HIVE MIND, WHICH DOES NOT EXIST AT ANU / PHOTON LEVEL. THE MOVIE IS DREAMED UP IN DARKNESS AIN'T IT? HOW DO YOU KNOW THAT FOR A FACT??? SIMPLE – ONE CAN NOT CONCEIVE SUFFERING IN A FIELD OF LOVE.

SO, HIVE MIND IS DREAMED INTO EXISTENCE BY OUR COLLECTIVE PURE CONSCIOUSNESSES EXISTING IN DARKNESS. THUS, PROVIDING FLAT 2D IMAGES THAT ARE SUCKED INTO THE DNA PART OF AN ANU VIA THE RULE = GRAVITY (AN AFFINITY FOR INFORMATION). THE DNA TAKES THE FLAT 2D IMAGES AND CONVERTS THEM TO A ROUND 2D HOLOGRAPHIC DIFFUSION PATTERNS VIA THE CROSS LINKS THAT MANIFEST YOU AND EVERYTHING ELSE, ACCORDING TO THE IMAGES / DREAM MOVIE PROVIDED BY HIVE MIND, WHICH INCLUDES EVERY THOUGHT THAT PASSES THRU YOUR HEAD. THE MOVIE INCLUDES EVERY GRAIN OF SAND ON THE BEACHES, TOO!!!

GENE EXPRESSION IS DUE TO COHERENT SUPER CONDUCTING LIGHT FLOWING IN THE SWITCHED-ON CHROMOSOMES – HOLOGRAPHIC PROJECTIONS ARE 3D LIGHT OBJECTS MAKING A FUZZY REPRODUCTIONS OF THE SUBJECT. SHINING A COHERENT LASER LIGHT THRU A 2D HOLOGRAPHIC DIFFUSION PATTERN CAUSES THE ORGANIZED RAYS OF THE LASER TO INTERSECT IN SPACE CREATING THE 3D LIGHT OBJECT. YOU DNA WORKS THE SAME WAY. AND, IT IS CONNECTED TO THE MOVIE.

THEN POLARITY TIGHTLY DEFINES THE "PHYSICAL" MATTER AROUND THE HOLOGRAPHIC PROJECTION – THUS, EVERYTHING YOU SEE, THAT HAS TIGHTLY DEFINED EDGES – IS DUE TO POLARITY / DUALITY – AND, IS MANIFEST BY A HIVE MIND WITH A PAST, PRESENT AND FUTURE, WHICH IS CHANGEABLE BECAUSE IT IS ALL MADE OF CONSCIOUSNESS

– 100th IDIOT EFFECT ALWAYS APPLIES AKA THE UNIVERSAL HOLOGRAM – THUS, ONLY COLLECTIVELY CAN WE CHANGE IT.

WHEN WE COLLECTIVELY BREAK OUT OF THE MATRIX, THEN INDIVIDUALLY WE WILL BE FREE!!!!

I AM HERE!!
I AM THE ONE
IN CHARGE OF THE FUN.
MY EUPHORIC WILL BE DONE!!! THANK YOU KINDLY!!!

SKID MARK DESIGN FEATURES:

BASICALLY, EVERYTHING YOU SEE IS AN ILLUSION / MAYA AS THE HINDUS CALL IT.

ALL OF NATURE IS A HOAX IN IT'S ENTIRETY – A CHICKEN TAUGHT ME THAT!!! IN HERE IS SCIENTIFIC PROOF. PLUS, FURTHER CONFIRMATION OF THE 3 SECOND SOUL MERGING PHENOMENON.
https://www.academia.edu/13078827/A_CONUNDRUM_-_THE_SOLUTION_IS_PROFOUND

THE ZODIAC IS A HOAX – DITTO HISTORY – THE LAWS OF PHYSICS ARE BOGUS

THE ROOT SOURCE OF VIOLENCE IS THE CARBON OXYGEN ANTAGONISM THAT EXISTS IN ALL CARBON-BASED LIFE FORMS. HOW ABOUT THIS FOR A DESIGN – THINKING – LIGHTS UP OUR NEURONS, THENCE THE SYNAPTIC GAPS CARRYING THE VIBRATION OF VIOLENCE. BECAUSE THE SYNAPTIC GAPS ARE ANALOG AMPLIFIERS AND CONNECTED TO THE 4-SIDED AETHERS, THEN THE INFO CROSSING THEM GETS BROADCAST. THEREFORE, THINKING MANIFESTS, AMPLIFIES AND BROADCASTS VIOLENCE. SOLID PROOF OF THAT IN THIS ESSAY:
https://www.academia.edu/34108091/THE_SOURCE_OF_VIOLENCE_IN_THIS_SOLAR_SYSTEM_Synaptic_bridges_help_brain_cells_communicate

JUDGMENT, DUALITY, FEAR, WAR, DEATH, SUFFERING, ETC. ARE ALL SKID MARK DESIGN FEATURES – REMEMBER, WE MAKE THEM EVERY MICROSECOND!!!!

THUS, IDIOTS UNITED CHURCH HAS BEEN FORMED, BECAUSE WE ARE ONE AND THUS YOU ARE A MEMBER. YOU JOINED UP SOME TIME AGO.
https://www.facebook.com/idiotsunitedchurch/?ref=profile_intro_card
INSPIRED BY THIS ESSAY
https://www.academia.edu/31014464/100th_IDIOT_EFFECT

SO, IT'S EASY TO HAVE COMPASSION FOR IDIOTS.
LET THE CELEBRATION BEGIN AND NEVER END

THE IDIOTS WIN AGAIN, IT IS OUR JOB.

THE HEAD IDIOT IS EXPOSED, SHOWS THAT YOURS TRULY IS LIVING OUT HIS KARMA. AT YOUR SERVICE - FOR ENTERTAINMENT PURPOSES ONLY
https://www.academia.edu/38539488/MERLIN_INDA_HOUSE_.docx   http://about.me/mike_emery

SUPPORTING DATA:

A LONG LIST OF SKID MARK DESIGN FEATURES IN HERE – LIKE THE SPOKEN WORD, THE BANE OF HUMANITY LEADS TO LIES, THENCE CRIME, POLITICS AND WAR. WHEREAS, TELEPATHY IS NORMAL AND WE DON'T DO IT!!! = A SKID MARK.
https://www.academia.edu/35457756/PROOF_OF_THE_MULTIVERSE_WAY_OF_THINGS.docx

https://www.academia.edu/35122597/CONSCIOUSNESS_BEGINS APPLIES THE PHYSICS OF CONSCIOUSNESS TO EDGAR CAYCE'S STATEMENT THAT ALL HUMAN CONSCIOUSNESSES STARTED AT THE SAME TIME.

WATCH MY VIDEO IT'S A BIT LENGTHY BUT FUN.
https://youtu.be/lq4z7A0qTes

THESE ESSAYS ARE NOT FOR THE MEEK. ONE OF MY GUIDES IS GEN. PATTON!!!
https://www.academia.edu/38628389/THE_ZODIAC_IS_A_HOAX
https://www.academia.edu/39034834/THE_LAWS_OF_PHYSICS_ARE_BOGUS
https://www.academia.edu/39613454/HISTORY_LESSON_No1

https://www.academia.edu/35597640/EVERYTHING_IS_A_HOAX_-
ALL_HUMANS_ARE_GODS_IF_EVERYTHING_OUTSIDE_OF_OUR_MINDS_IS_A_HOAX_ILLUSION_AND_IT_IS_THEN_US_TALKING_HUMANS_CREATE_IT_EVERY_MICROSECOND_AND_WE_DO._WE_PLANTED_THE_HOAX_SO_THAT_WE_WILL_WAKE_UP_THIS_TIME_DUE_TO_THE_INTERNET

https://www.academia.edu/33280902/EARTH_IS_DYING_-_THE_IDIOTS_ARE_WATCHING DONE FROM THE STAND POINT THAT WE COLLECTIVELY ARE GOD AND WE ARE DOING THIS TO OURSELVES

ECKHART TOLLE: https://youtu.be/rWFVi1cPUZo?t=247 SAME VIDEO AS ABOVE--> IT IS NOT PERSONAL!!! HAHAHAHA HEHEHE HAHAHA

SO, **EUPHORIA!!!! IS THE SOLUTION – MORE IN "HOW TO BREAK OUT OF THE MATRIX (2)"**

WHEN USING A EUPHORIC ORGANMIC BUBBLE – THE 100th IDIOT NUMBER REQUIRED TO ACHIEVE A QUANTUM LEAP GOES WAY DOWN. https://www.academia.edu/40198090/ITS_RAINING_IN_THE_AMAZON_-_MAJOR_BUBBLE_TECH_WIN
WE MADE EUPHORIC RAIN IN THE AMAZON, SIBERIA, THE CONGO AND AUSTRALIA. THIS AUSSIE REPORTS ARE AWESOME – NO QUESTION THAT BUBBLE TECH IMPOSED RAIN AND TURNED FIRES AROUND. https://www.academia.edu/41826736/WINNING_IN_AUSTRALIA_WITH_RAIN_AND_COOLER_WEATHER

ACID QUESTION - ARE YOU PART OF THE SOLUTION OR ARTWORK IN THE PROBLEM?? SUFFICE TO SAY, THAT IF YOU ARE NOT USING YOUR CONSCIOUSNESS TO WILLFULLY TAKE FUNCTIONAL CONTROL OF THE ENVIRONMENT, THEN THE JUDGMENTAL DUNG IN YOUR MIND MAKES YOU PART OF THE PROBLEM – A BIG PART. LOOKING AND LABELING = IDIOTS.
https://www.academia.edu/38628283/RED_PILL_BLUE_PILL

YAHOOOOOO

IT WILL APPEAR FUZZY TO US – DREAMY!!!

Earth's Epochs, 2017

John Jensen

In this book, I am expecting to prove via catastrophic ocean rise, the geologic column, deep core drills, ice core samples, varve layers, and other readings, that the Earth experiences major and cataclysmic catastrophes including dipole exchange on an irregular short-term basis.
https://www.academia.edu/38408870/Earth_Epochs_full_book_in_pdf_format?email_work_card=view-paper

INTERESTING THAT SCIENCE FINDS IRREGULAR PERIODS OF CALM / BLANK IMAGINEERING VS THE REGULAR CYCLES AS DETAILED IN THE ZODIAC, MAYAN CALENDAR AND HINDU YUGAS ARE THEN HOAXES - YIKES. MAKING THE BLINKING IN OF ALL OF THE SAME DNA EVEN MORE AMAZING. WE GO

TO ANY LENGTH TO PLAY THE OLD MOVIE – WE COLLECTIVELY ARE TOO LAZY TO CHANGE IT.

THEN ONE ASSUMES THAT THERE ARE DIVINE BEINGS, WHO DO HAVE THE ABILITY TO CHANGE LARGE AREAS AND TEACH US TELEPATHY, BUT THEY DON'T = HIVE MIND HEROES OF HIERARCHY = ALL DIVINE BEINGS, SO FAR, ARE B.S. SORRY.

WE HAVE ARRIVED AT DO-IT-YOURSELF TIME – SURPRISE IS SUPREME, THE HEIGHT OF IMAGINATION IS EUPHORIC SURPRISE.!!! LET THE CELEBRATION BEGIN AND NEVER END!!!

FURTHER PROOF THAT IT IS THAT MERLIN IS DOING THIS AND NOT THIS PHYSICAL IDIOT

MIKE'S THEORY OF EVERYTHING

VIDEO https://youtu.be/2lgoKI9MyM0

IT'S NOT A THEORY, JUST A STATEMENT OF THE FACTS. WRITTEN ABOUT 20 YEARS AGO
https://www.academia.edu/11138575/MIKES_THEORY_OF_EVERYTHING

THIS PAPER HAS BEEN TESTED BY KINESIOLOGY AND DOWSING AND FOUND TO BE THE BEST-KNOWN EXPLANATION OF HOW THIS UNIVERSE IS CONSTRUCTED.

THIS PHILOSOPHY ABOVE IS ACTUALLY PHYSICS THAT STEMS FROM MIKE'S THEORY OF EVERYTHING I.E. **IT ALL LINKS OR DOVE TAILS PERFECTLY FROM THE VOID VIA THE RULE TO THE INTENTIONAL MANIFESTATION OF THIS IDIOTSVILLE.**

MIGHT BE A HUGE MESSAGE???

THE MATRIX (2) – BREAK OUT YOU MUST!!!

TIME AND SPACE DO NOT EXIST BECAUSE, THE HUMAN BODY HAS INFINITE AWARENESS.

LET YOUR BODY FEEL YOUR WAY OUT OF THE MATRIX – TRUST IT. IT'S TIME FOR THE PUPPETS TO USURP THE REINS OF REALITY!!

BECAUSE, AS YOU ARE ABOUT TO LEARN VIA MULTIPLE DATA POINTS: **SPACE AND TIME DO NOT EXIST IN THE WAY THAT YOU THINK THEY DO. THUS, EVERYTHING IS INSIDE OF YOU NOW!!!** FIRSTLY, THOU, WHOOPS. ---->

THE IDIOT IS EASY TO FIND – HE THINKS IDIOTSVILLE IS REAL AND WILL TELL YOU SO!!! HAHA HAHA

THE FIRST STEP TOWARD FREEDOM IS TO BECOME AWARE OF THE MATRIX AND YOUR PUPPETHOOD THEREIN!!!

IT GETS WORSE, SEE BELOW:

THIS IS IRREFUTABLE: THE PUPPET, HUMAN BODY, HAS INFINITE AWARENESS, QUITE LITERALLY EVERYTHING KNOWS EVERYTHING, IT'S ALL CONNECTED TO THE MATRIX

MOVIE AND YOU DON'T KNOW IT, BUT YOU ARE ABOUT TO FIND OUT.

OUR BODY'S INFINITE AWARENESS IS NOT TO THE EXTENT OF OUR PURE CONSCIOUSNESS BECAUSE OUR MEMORIES ARE STORED IN STRUCTURES / ANUs THAT HAVE A TIME COMPONENT AND DO OCCUPY A BIT OF SPACE WITHIN OUR & GOD'S PURE CONSCIOUSNESSES. IT'S RIGHT BEHIND YOUR EYES, WITHIN YOUR MIND AND NO WHERE ELSE. AMAZINGLY, THE ENTIRETY OF CREATION INCLUDES A LARGE NUMBER OF MULTIVERSES THAT ARE INSIDE OF OUR PURE CONSCIOUSNESSES.
https://www.academia.edu/38629357/THE_EXPANSIVENESS_OF_OUR_CONSCIOUSNESSES

AS PURE CONSCIOUSNESSES, WE PROBABLY HAVE MORE AWARENESS THAN GOD HERSELF – THE TELLTALE SIGN OF THAT, IS THAT SHE MISSES, IS NOT AWARE OF, ENTIRE MULTIVERSES THAT MIRROR OFF WITHIN HER SPACE AS REPORTED IN THE EXCELLENT BOOK, "CONVERSATIONS WITH GOD". IT SEEMS THAT SHE ONLY HAS TWO MODES – THE ULTIMATE LONELY OBSERVER (THE OLD BITCH) OR OUR LOVING CREATOR MOTHER WHO IS IN CONTINUOUS ORGASM, WHICH WE, AS PURE CONSCIOUSNESSES, HAVE BROKEN AWAY FROM – AS A TRIBE. THE 100th IDIOT EFFECT ALWAYS APPLIES. THUS, OUR PURE CONSCIOUSNESSES ARE NOT AFFECTED BY THE CONTINUOUS ORGASM THAT GOES ON, THENCE MORE EXPANSIVE THAN HERS!!!

WE HAVE BEEN WORKING ON THIS PROJECT FOR AN UNKNOWABLE LENGTH OF TIME – EUPHORIA IS A CHANGE TO THE ENTIRETY OF CREATION. AND YOU ARE HERE TO FEEL IT AND SEE THE RESULTS. THE PROMISED GOLDEN AGE IS VERY SOON UPON US.

THE EVOLUTION OF EMOTIONS:

THE MATRIX IS FUELED BY MEMORIES BECAUSE THE PAST PORTENDS A FUTURE. ALL OF THOSE MEMORIES ARE BOUND TOGETHER AS A SERIES OF IMAGES BY EMOTIONS, WHICH UP

UNTIL NOW HAVE ALWAYS BEEN EITHER LOVE OR FEAR BASED. EUPHORIA SUPPLANTS BOTH LOVE AND FEAR AND THUS, RELEASES THE BONDS THAT HOLD THE DUALITY BASED MEMORIES TOGETHER, WHICH EXPUNGES THOSE MEMORIES. FEAR BASED IDEAS WILL BECOME COGNITIVE DISSONANCE DITTO LOVE BASED TO A CERTAIN EXTENT BEST TO BE CONVERTED TO ELATION, WHICH WE CAN DO BY FORCE OF OUR WILL. YAHOOOOOO!!!

MERLIN WAS ON TO THIS SOME TIME AGO!!!
https://www.academia.edu/24643924/THE_EVOLUTION_OF_FEELINGS

EUPHORIA, ALSO, SOLVES THE PROBLEM OF GOD'S LONELINESS AND THE DARKNESS BECOMES NOT DUALISTIC, NOT DARK AND NOT LIGHT THUS THE EUPHORIC GRAY IN VARYING SHADES IS OR HAS BECOME THE NEW BLANK SUBSTRATE OF CREATION REPLACING DARKNESS. MORE ON THAT IN THE MATRIX (3).

LOVE:

THE LOVEY DOVEY, METAPHYSICAL, FLUFF HEADS DON'T GET IT – IF YOU MERGE WITH LOVE, GETTING OUT IS DIFFICULT. AND **IT PERPETUATES THE ABORIGINAL DUALITY OF LOVE AND FEAR** – BLINKING IDIOTS. WELL, IF A CONTINUOUS STREAM OF LOVE INDUCED ORGASMS WORKED AS THE BASIS OF A LONG-TERM REALITY, YOU'D BE IN IT. BUT, YOU ARE NOT BECAUSE IT'S BORING. I WONDER HOW LONG IT TOOK FOR THAT BOREDOM TO SINK IN?? WHY DO YOU THINK THAT WE RELIGIOUSLY LEGISLATE AGAINST LOVE AND ORGASM AS BEING THE "ORIGINAL SIN"? HUH?

ANYWAY, IT GETS WORSE

THE HUMAN BODY HAS INFINITE AWARENESS AS DOES EVERYTHING ELSE. AND, YOU ARE THE BUTT OF THE JOKE. IT'S NOT PERSONAL. THE SHOW MUST GO ON.

1) KINESIOLOGY: 10 YEARS, 10 AMERICAN UNIVERSITIES, 20 MILLION MUSCLE TESTS LATER ARE ALL SUMMED UP WITH ONE SENTENCE = THE HUMAN BODY WILL ANSWER PERFECTLY ANY YES OR NO QUESTION ON ANY SUBJECT. THAT'S INFINITE AWARENESS. WHY? SEE 2)
http://blog.hasslberger.com/docs/KINESIOLOGY-HAWKINS.pdf

2) BECAUSE: LIKE ALL THINGS, THE HUMAN DNA / ANU IS CONNECTED TO IT'S HIVE MIND IN THE FORM OF IT'S PAST, PRESENT AND FUTURE, ALL OF THE TIME. ACQUIRING THE KNOWLEDGE OF ONE'S PAST, PRESENT AND FUTURE IS INVOLUNTARY, AS THE RUSSIANS PUT IT. THIS AS A RESULT OF DECADES OF WORK WITH CYLINDRICAL MIRRORS, KNOWN AS THE KOZYREV MIRRORS. PUT YOUR BODY IN A PROPERLY CONSTRUCTED CYLINDRICAL MIRROR AND YOU WILL SEE YOUR PAST, PRESENT AND FUTURE, WHETHER YOU WANT TO OR NOT. NOSTRADAMUS SAT INSIDE OF AN EGG-SHAPED DEVICE WITH A SILVER MIRROR FINISH AND HE SAW THE PPF OF THE PLANET. OUR BODIES HAVE INFINITE AWARENESS OF THE MANIFEST UNIVERSE I.E. EVERYTHING MADE BY AN ANU – WHAT DO YOU NOT UNDERSTAND ABOUT GOD?

AND, AND, WE CAN ACCESS THE PAST, PRESENT AND FUTURE OF ANYTHING WE LOOK AT FOR 3 SECONDS WHILE IN OUR SPIRITUAL BODY, THERE COMES A SEAMLESS MERGING – THEN NO, YOU OR I – IT'S ALL ONE MOVIE. THIS OCCURS IN THE RUSSIAN MIRRORS. AND ALSO, VIA LUCID DREAMING, ASTRAL PROJECTION, NDE AND DEEP MEDITATION. THIS IS NORMAL CONSCIOUSNESS DUE TO THE RULE, LIKE ATTRACTS LIKE, WHICH IS DIFFICULT TO ESCAPE FROM. EUPHORIA SOLVES THE ALL-KNOWING PROBLEM, TOO. FUZZES THINGS OUT – NO HARD EDGES – DREAM LIKE.

HERE'S AN EXCELLENT ESSAY PROVING THE POINT–
https://www.academia.edu/13078827/A_CONUNDRUM_-_THE_SOLUTION_IS_PROFOUND

THE CONUNDRUM WAS THIS: FOR A HUMAN'S DNA TO STORE AND BE ABLE TO PROJECT ONTO A MIRROR, THE

BODY'S PAST, PRESENT AND FUTURE (PPF) THIS IS NOT DIFFICULT TO UNDERSTAND, HOPEFULLY – WELL, FOR 1000s OF YEARS THE HINDUS HAVE KEPT THE AKASHIC RECORDS, WHICH HAVE THE PPF OF EVERYTHING. SO, HOW IS IT THAT WE HAVE ACCESS TO THE PPF OF EVERYTHING?? THE SOLUTION IS THE MIND OF GOD WHICH WE NOW KNOW IS A HIVE MIND – EVERYTHING IN THE HIVE MIND MOVIE IS CONNECTED TO IT ALL!!! THAT ELIMINATES TIME AND SPACE, COLLAPSING IT INTO THE SEED OF THE ENTIRE MULTIVERSE, WHICH IS INSIDE OF YOU AND NOWHERE ELSE. WELL, CAN YOU IMAGINE – WHERE WOULD YOU STORE ALL OF THAT INFORMATION?? QUITE SIMPLY IF IT WAS ANYWHERE OUTSIDE OF YOU, THEN YOU'D SEE IT. EH?

PLEASE, VIA MICROSCOPES AND TELESCOPES WE HAVE LOOKED AT EVERYTHING AND NOT FOUND OUR PPF OUTSIDE OF OURSELVES. THEREFORE, MY FINE FEATHERED IDIOT IT'S INSIDE OF YOU. HAPPILY, OUR HEART CHAKRAS ARE ACTING AS OUR GUIDES OUT OF THE MATRIX. EXCELLENT ESSAY ABOUT OUR HEART GENIES:
https://www.academia.edu/37948640/BUBBLE_GENIES_ARE_LEADING_US_TO_FREEDOM_.docx

JUST ABOUT EVERYBODY HAS HAD A PRECOGNITIVE DREAM OR DE JA VU. WHY?? THAT IS A SMALL WONDER GIVEN THAT THE ENTIRE PAST, PRESENT AND FUTURE IS AVAILABLE. AND IT IS ALL CONNECTED, IT MUST MARCH IN LOCKSTEP JUST LIKE A PICTURE PUZZLE, WHICH = HIVE MIND – DON'T IT? WELL, HOW IS IT THAT ALL OF THESE DREAMED SEQUENCES OF EVENTS FIT TOGETHER, IF IT IS NOT ONE MOVIE.

3) THE MAX PLANCK INSTITUTE – THE HUMAN BODY IS AWARE OF WHAT THE MIND IS GOING TO DECIDE UP TO 7 SECONDS BEFORE HAND.. THIS MUST BE A CONTINUOUS STATE OF PRECOGNITION. SEE SUPPORTING DATA BELOW:

SO, IN COMES A HAND GRENADE, THAT THE BODY KNOWS IS COMING AND OUR MEEK IDIOT BODY WAITS TO BE BLOWN UP. THIS APPLIES TO ALL DEATH EVENTS.

WHO'S THE BLINKING IDIOT??? THE WORD HUMANITY DOES NOT APPLY. WE HAVE A HIVE MIND DREAMED UP BY OUR PURE CONSCIOUSNESSES THAT HAS CREATED A LONG-PLAYING MOVIE BASED ON DUALITY FOR THE PURPOSE OF ENTERTAINMENT VIA SEPARATION – AND, IT'S A MASSIVE GAME THAT GOES ON INSIDE OF OUR MINDS.

AS DR. MOSHEN SAYS ON THE MATRIX PANEL ABOVE, "to study emotions and achieve expanded consciousness." WE AS PURE CONSCIOUSNESSES HAVE IMPOSED THE EXTREMES OF LOVE AND FEAR ON A LIVING SYSTEM FOR CREATIVITY PURPOSES. BECAUSE, ALL KNOWING HAS NEXT TO ZERO CREATIVITY, DITTO CONTINUOUS ORGASM, WHICH IS WHY WE ARE HERE. IT'S NOT PERSONAL, IT'S JUST ENTERTAINMENT.

THUS, CLEARLY EUPHORIA IS THE NEXT EVOLUTION IN THE ROOT EMOTIONS OF CREATION.

THE CIA WORKED WITH THE MONROE INSTITUTE TO DEVELOPED A TECHNIQUE FOR MANIFESTATION, CALL THE GATEWAY PROCESS. THIS VIDEO IS A DECLASSED DISCLOSURE OF THAT, REPORT READ ALOUD, https://youtu.be/YalTy-H-Tjk?t=6134 THE LINK TO THE WRITTEN REPORT IS IN THE CREDITS. THE CRUX OF WHICH IS TO MEDITATE IN THE NOW.

Eckhart Tolle Says

As you become comfortable with uncertainty, infinite possibilities open up in your life. It means fear is no longer a dominant factor in what you do and no longer prevents you from taking action to initiate change.

"WITH THE INTENTION THAT THE DESIRED OBJECTIVE IS ALREADY AN ESTABLISHED ACHIEVEMENT, WHICH IS DESTINED TO BE REALIZED WITHIN THE TIME FRAME SPECIFIED" = GODHOOD MIND / OVER MATTER IF IT IS IN HARMONY WITH THE "UNIVERSAL HOLOGRAM", AS THEY PUT IT, WHICH IS HIVE MIND, THAT THEY DID NOT KNOW ABOUT BACK THEN. BREAKING OUT MEANS NO THOUGHTS, JUST FOLLOW YOUR FEELINGS!!!

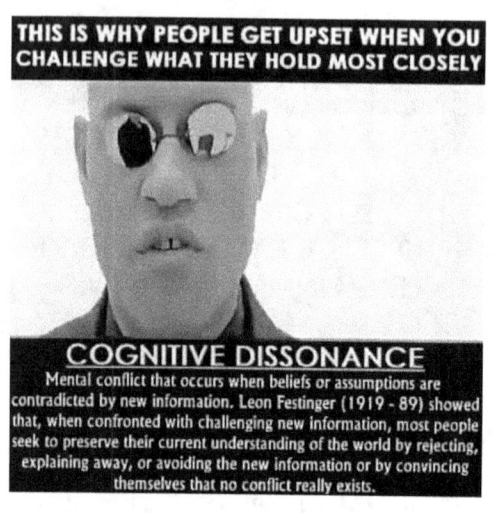

COGNITIVE DISSONANCE IS GOOD IN SPIRIT, SURPRISE IS SUPREME – IDIOTS WIN AGAIN

S0M3T1M3Z WE N33D 2 DIS-C0NNECT 2 C0NNECT

OPERATING IN A FUZZY ENVIRONMENT REQUIRES NO FEAR. YOU DON'T WANT TO KNOW WHAT IS GOING TO HAPPEN. ONLY HOW YOU FEEL ABOUT WHAT IS HAPPENING. THENCE, LEARNING TO APPLY YOUR WILL IN THE FUZZY ENVIRONMENT.

THE NEXT DATA POINT TELEPORTATON:

WE TELEPORT IN SPIRIT, EVERYTHING DOES, ALL OF THE TIME, IT'S A BLINKING SHOW AT THE ROOT LEVEL THAT YOU CAN SEE – THIS

IMAGE CAPTURED BY OAK RIDGE NATIONAL LABS SHOWS WATER MIRRORING OFF INSIDE OF HEXAGONAL TUBULES / "channels"
https://phys.org/news/2016-04-state-molecule.html#jCp ""This means that the oxygen and hydrogen atoms of the water molecule are 'delocalized' and therefore simultaneously present in all six symmetrically equivalent positions in the channel at the same time." MEANING THAT THE WATER MIRRORS OFF ON ALL SIX WALLS OF THE HEXAGONAL AETHERIC CELL, THEN IT JUST APPEARS IN THE NEXT AETHERIC CELL – APPEARING TO BE A SMOOTH MOVE, BUT IT BLINKS.

THESE ARE THE MICROTUBULES THAT MIT OBSERVED TO EXIST IN ALL THINGS. THEY ARE PARALLEL AND HAVE WATER MOLECULES AS THE CROSS LINKS – SO, TWO SEPARATE OBSERVATIONS SEEING THE SAME THING. DESCRIBING IT DIFFERENTLY, OF COURSE = IDIOTSVILLE = OWNERSHIP OF INFORMATION WHICH = SURVIVAL = IDIOTSVILLE. HAVE YOU EVER WORRIED ABOUT MONEY?
http://blog.hasslberger.com/docs/MICROTUBULES.pdf

THUS:

https://www.academia.edu/33528642/TELEPORTATION_IS_NORMAL.doc

THE MOVIE IS INSTALLED IN THE BLINKING GAP, THAT IS NOW EUPHORIC AND IT USED TO BE LONELY. THIS EUPHORIC EMOTION IS A DESIGN RECENTLY INSTALLED, SINCE REICH'S TIME, BECAUSE WITH TOO MANY DEVICES AROUND REICH'S LAB, HE AND HIS LAB ASSISTANTS USED TO EXPERIENCE ORGONE POISONING – VS – VIA THE EUPHORIC BUBBLE TECH THAT DOES NOT HAPPEN, IT IS ALL JUST WONDERFUL EUPHORIA. THERE IS NO DEADLY ORGONE RADIATION (DOR) PRODUCED THAT WE HAVE BEEN ABLE TO FIND – WHEREAS, NORMAL ORGONE DEVICES GENERALLY HAVE A DOR REGION TO THEM.

AND IT IS HIGHLY LIKELY THAT REICH HIMSELF CONTINUED TO WORK ON THE ORGONE POISONING PROBLEM IN SPIRIT AND HE MAY HAVE FIGURED THAT OUT – HHHMMMMM YUP.

ALMOST WITH CERTAINTY WILHELM REICH IS GUIDING THIS PROJECT – HE DID START IT!!!

SO, NOWADAYS, REICH'S TECH AND VIRTUALLY ALL DEVICES BASED ON CAPACITANCE EMIT EUPHORIA – THIS IS AN INSTALLED DESIGN BY THE HIVE MIND OF OUR COLLECTIVE HUMAN BODIES, AS WELL AS, THE BODIES OF THE WILD ANIMALS THAT DO DREAM AND WANT TO LIVE NOT DIE. WE ARE STILL IN A HIVE MIND AS MADE OBVIOUS BY THE HARD EDGES OF THE "REALITY" THAT YOU SEE. SUCH EDGES ARE DUE TO POLARITY WHICH IS DUALITY WITH THE ATTENDANT MOVIE RUNNING THE SHOW.

SO, INSTANT TELEPORTATION CAN ONLY OCCUR IF THERE IS NO TIME OR SPACE - SNICKER. LEAVING THE LOCATION OF REALITY AS BEING ONLY IN YOUR MIND.
https://www.academia.edu/34247198/WHAT_DO_YOU_SEE_.doc

ANSWER: WHAT YOU SEE WHEN YOU LOOK OUT IS WHAT IS GOING ON INSIDE OF YOUR MIND, WHICH IS THE HIVE MIND MOVIE.

THE AMAZING PART IS THAT WE ESCAPE THE MATRIX TO A CERTAIN EXTENT VIA DREAMING. ONE CAN GO ANYWHERE IN SPACE AND TIME INSTANTLY. WE ALL DO IT EVERY NIGHT WHEN WE DREAM. AND VIA FORCE OF OUR WILL IN LUCID DREAMING, ASTRAL PROJECTION, NDE, REMOTE VIEWING AND DEEP MEDITATION WE CAN TELEPORT WILLFULLY SPIRITUALLY.

THEN VIA THE TESLA UTRON (LOOK IT UP) PEOPLE HAVE TELEPORTED PHYSICALLY. AND AFTER ENOUGH RIDES ON THE UTRON, THE DEVICE IS NO LONGER NEEDED - WE TELEPORT ALL OF THE TIME ANYWAY. SO, YOU CAN GO ON YOUR OWN VIA YOUR OWN WILL TO DO SO. BE CAREFUL WHAT YOU WISH FOR.

THIS IS A TECHNOLOGY CONSCIOUSNESS INTERFACE, THE TESLA UTRON HAS NO MANUEL CONTROLS. IT COMPLIES

WITH YOUR WISHES, LIKE THE DISNEY MOVIE, "FLIGHT OF THE NAVIGATOR" ONLY THE UTRON GETS THERE INSTANTLY.

IF EVERYBODY KNEW, UNDERSTOOD AND USED WHAT FOLLOWS, THEN WE CAN DO IT OURSELVES DIY VIA FORCE OF OUR OWN WILL. AND NOW, TO HELP WITH THAT KIND OF TRAVEL WE HAVE THE EUPHORIC BUBBLE TECH THAT CLEANS THE AGRO / LOOSH / SEPARATION ENERGY OUT OF THE AETHERS SO THAT WE CAN TELEPORT AND TELEPATH EASIER. LOOSH IS A TERM COINED BY ROBERT MONROE MEANING SEPARATION ENERGY.

MAJOR QUESTIONS ABOUT TELEPORTATION:

AS YOU CAN SEE, EVERYTHING MIRRORS OFF ACROSS A GAP = TELEPORTATION.

HOW DO YOU GET THERE INSTANTLY?? IN FACT, VIA SPIRIT TRAVEL, YOU GET THERE BEFORE YOU FINISH ASKING THE QUESTION – THIS IS THE WAY THAT YOU KNOW IT IS A REAL SPIRITUAL EVENT. THE REASON THAT YOU GET THERE INSTANTLY IS BECAUSE YOU ARE ALREADY THERE AS YOUR AURA / ANU - THE GOD PARTICLE THAT IS OMNIPRESENT. BI-LOCATION IS POSSIBLE TOO.

SINCE WE CAN GO ANY WHERE IN SPACE AND TIME INSTANTLY, BECAUSE OUR ANU IS ALREADY THERE – THEN OUR ANU AND US EXIST EVERYWHERE IN SPACE AND TIME. SAID ANOTHER WAY: SINCE WE CAN GO ANYWHERE IN SPACE AND TIME INSTANTLY BECAUSE OUR ANU IS ALREADY THERE – THEN **OUR ANU HAS TO BE EVERYWHERE NOW,** WHICH AGAIN MEANS THAT SPACE AND TIME DO NOT EXIST!!!!

NEXT A QUESTION, THE PHYSICS THEREOF: WHO MADE THE ANU THERE??? WHOOPS!!! SINCE YOU ARE THE REAL THING, THEN THERE IS A REFLECTION OF YOU. AIN'T IT? **THUS, EVERYWHERE IS A REFLECTION OF YOU!!!** SO, AGAIN:

THERE IS NO THERE,
THERE IS ONLY HERE AND NOW
ALL REALITY IS WHERE YOU ARE RIGHT NOW

WHICH MEANS THAT SPACE AND TIME ARE A FIGMENT OF OUR COLLECTIVE IMAGINEERING.

AND THUS, CLEARLY IT IS ALL A MOVIE GOING ON IN OUR SUBCONSCIOUS = THE HIVE MIND MATRIX.

WHEN WE BREAK OUT OF HIVE MIND, THIS WILL BE THE CASE:

I AM THE ONE
IN CHARGE OF THE FUN
MY EUPHORIC WILL BE DONE!!!
NOW!!!, SO-BE-IT
THANK YOU KINDLY - THIS YOU CAN DO IN SPIRITUAL TRAVELING NOW!!! SEE TELEPATHY TECH BELOW.

KNOWING THAT AND USING THE FEELINGS OF OUR FINE 3D HUMAN BODIES, WE CAN BREAK OUT OF THE MATRIX ANY TIME, CYCLES NOT DEPENDENT ONLY OUR COLLECTIVE WILL IS REQUIRED. THE 100th IDIOT EFFECT ALWAYS APPLIES.

ORGANMIC BUBBLE TECH ALLOWS ALL HUMANS TO INSTALL THEIR DESIRE FOR FREEDOM AT SEED LEVEL, WHICH IS WHERE IT COMES FROM REALLY, OUR ANUs / SEEDS ARE CONSCIOUS OF BEING TRAPPED IN AETHERIC CAGES. THEREIN LIES THE ROOT DESIRE FOR FREEDOM.

OUR FORCE OF WILL IS SUPREME IN SPIRIT

https://www.academia.edu/32877101/TELEPATHY_TECH_IS_TRAINING_WHEELS_FOR_GODS_.doc

- - **SO, IT'S EASY TO HAVE COMPASSION FOR IDIOTS.**
- - **LET THE CELEBRATION BEGIN AND NEVER END**
- - **THE IDIOTS WIN AGAIN; IT IS OUR JOB.**

Gautama Buddha Says

The fool who thinks he is wise is just a fool. The fool who knows he is a fool is wise indeed.

THE HEAD IDIOT IS EXPOSED, SHOWS THAT YOURS TRULY IS LIVING OUT HIS KARMA.
AT YOUR SERVICE - FOR ENTERTAINMENT PURPOSES ONLY – MERLIN.
https://www.academia.edu/38539488/MERLIN_INDA_HOUSE_.docx   http://about.me/mike_emery

SUPPORTING DATA: Kozyrev's mirror

Imagine standing under a vast, scintillating Aurora Borealis, and seeing it change colors as you changed your thoughts. This exact situation led Russian medical doctor Alexander V. Trofimov into his groundbreaking research on human consciousness, in collaboration with Vlail P. Kaznacheev, and following in the footsteps of the great 20th century physicist Nikolai Kozyrev.

Essentially, Kozyrev devised reproducible experiments that prove the existence of a "torsional energy field" IT APPEARS TO BE A TORSIONAL FIELD BECAUSE THE AETHERS ARE TWISTED A BIT BY THE INFO TRAVERSING THEM beyond electromagnetism and gravity, which travels much faster than light. He called it the "flow of time." Others, Einstein among them, have called it "ether." Others call it "zero-point energy." IN OTHER WORDS, KNOWING EXACTLY HOW THE UNIVERSE IS CONSTRUCTED IT USEFUL -

https://www.academia.edu/37978948/EXACTLY_HOW_THE_UNIVERSE_IS_CONSTRUCTED.docx

Within this "flow of time," the past, present, and future all exist at the same time, and in every place. This discovery sets the stage for all psychic phenomena to be scientifically explainable. Trofimov and Kaznacheev have, for the past thirty years, been experimentally developing the practical explanations, and have made some surprising discoveries.

The first, dubbed "Kozyrev's Mirrors," reflects thought energy (which exists within the "flow of time") back to the thinker. This apparatus, invented by Kozyrev, gives access to intensified consciousness and altered states, including non-linear time — similar to a deep meditational state.

Trofimov's work has consisted of "remote viewing" experiments across both distance and time. They discovered that results are more positive when the "sender" is in the far north, where the electromagnetic field is less powerful. **So, they invented a second apparatus that shields an experimental subject from the local electromagnetic field. Within this apparatus, their subjects can reliably access all place and time — past, present, and future — instantaneously.** Construction specifications for these apparati are published in Russian scientific literature.

Among Trofimov and Kaznacheev's conclusions are:
1) our planet's electromagnetic field is actually the "veil" which filters time and place down to our everyday Newtonian reality — enabling us to have the human experience of linear time,
2) in the absence of an electromagnetic field, we have access to an energy field of "instantaneous locality" that underlies our reality,
3) that the limiting effect of the **3D** electromagnetic field on an individual is moderated by the amount of solar electromagnetic activity occurring while that person was in utero, and

4) that once a person has accessed these states, his or her consciousness remains so enhanced.

The implication is that the global electromagnetic soup of cell phones, radio, television and electric appliances actually impedes our innate communication abilities. The further implication is that expanded human consciousness is mechanically producible now, which raises the vast ethical question of how these apparati can be most beneficially used.
Complete Interview here:
https://quantumpranx.wordpress.com/?fbclid=IwAR3exk5z1Rheiu-6ZAWZa5hqEnED7lVkTjinnXa-K_Ycu6Ovgfqav_zoHL8

Kozyrev mirrors were used in experiments related to extrasensory perception (ESP), conducted in the Institute of Experimental Medicine of Siberia, division of the Russian Academy of Sciences. Humans, allocated into the cylindrical spirals (usually 1.5 rotations clockwise, made of polished aluminum) allegedly experienced anomalous psycho-physical sensations, which had been recorded in the minutes of the research experiments.

A Kozyrev mirrors is a device made from aluminum (sometimes from glass, or reflecting mirror-like material) spiral shape surfaces, which are able to focus different types of radiation including that coming from biological objects.[citation needed] They are named after the famous astronomer Nikolai Aleksandrovich Kozyrev, though they were neither invented nor described by him.
http://kozyrevmirrors.com/kozyrev-mirrors/?fbclid=IwAR0X24KWKRaSpSQLR_BldUnVtDKkVVSdWxfZW15vCPzg42Lxk9eyHik8SA

A 1998 Russian patent, RU2122446, "Device for the correction of man's psychosomatic diseases", relates to Kozyrev mirrors.

The mirror of Kozyrev is an invention by astrophysicist Kozyrev. The mirror of Kozyrev is a cylinder, made of aluminum, which opens a time-space channel by its inherent size, surface conditioning and placement. The assertion is confirmed by numerous scientific experiments with hundreds of volunteers and by the scientific theory of one of the greatest astrophysicists of the last century, Nikolay Kozyrev.

Experiments with the mirror of Nikolay Kozyrev
The experiments of Russian scientists are performed in the nineties and their results are quite interesting. Based upon Kozyrev's findings they built an aluminum cylinder, since this cylinder is made by materials that reflect the Time Radiation for 100%. This way, it should be possible for

people in a Kozyrev mirror to receive pure mental information from other areas of the earth and the universe.

The research is done by the astrophysicist Kozyrev and Russian scientists. Knowledge was discovered of modern biology about the substance dimethyltryptamine, the functioning of the epiphysis, different cultures, as the Egyptian culture during the periods of the Pharaohs, notes on the Ka-body and the so-called etheric double.

We know that the epiphysis produces dimethyltryptamine (DMT) from melatonin. DMT is a hallucinating substance. This is called hallucinating, because the reality people are describing under the influence of this substance, is experienced to be an illusion. These experiences don't fit in the western line of reasoning of the reality and thus they are rejected. In the mirror of Kozyrev these story lines can come together. The mirror is a cylinder that is as high as a grown-up person and you enter it. Inside this mirror is a special (torsion) field, with a different time space, the field stimulates the epiphysis and communication with the higher worlds is possible. Research has found the possibility of a profound power of telepathy. The stimulation of the mirror on the epiphysis enhances the silver cord. In this process we experience the Ka-body. The Egyptians have all sorts of exercises to do this, as their goal is to send energy to the Ka-body. The Egyptians were searching for eternal life, like the Chinese and the alchemists in the west. The mirror is capable of giving us certain insights in our own life and help finding more vitality/health. The Kozyrev mirror we're using is completely different from the existing Kozyrev mirrors. Our Kozyrev mirror is connected to frequency equipment and is therefore creating a much deeper effect than with the already existing Kozyrev mirror. The mirror works both spiritual and physical and gives many people a tingling and vitalizing effect. This mirror enables us to release our consciousness of strict beliefs we learned ourselves, or were taught by our parents, grandparents, ancestors, schools, media etc. The mirror can broaden our horizon, a new framework of references, making us free from rigid patterns/thoughts. This is capable of giving us a completely new life with totally different perceptions. Meaning that old stress-patterns can be cleaned out and as a result we start to feel better/more vital and this again will reflect in our daily lives and situations at home and at work, in a much more positive course of everything. The mirror has different influences on each of us, because we all have our own specific frequency (vibration of our cells). Every human being is unique. The effects may soon be visible for one person, while it may take a while for someone else. The same goes for the feeling of perception. Some people notice and feel something after just one session at the mirror and another person notices something after multiple times. The mirror can act as a gigantic spiritual, physical and emotional life-rollercoaster. The mirror can make us feel younger, better, happier and extremely good.

FROM THE MAX PLANCK INSTITUTE WEBSITE:

https://www.mpg.de/research/unconscious-decisions-in-the-brain

THEY SHOULD KNOW, NOTHING IS UNCONSCIOUS.

Unconscious decisions in the brain

A team of scientists has unraveled how the brain unconsciously prepares our decisions

APRIL 14, 2008

Brain

Already several seconds before we consciously make a decision its outcome can be predicted from unconscious activity in the brain. This is shown in a study by scientists from the Max Planck Institute for Human Cognitive and Brain Sciences in Leipzig, in collaboration with the Charité University Hospital and the Bernstein Center for Computational Neuroscience in Berlin. The researchers from the group of Professor John-Dylan Haynes used a brain scanner to investigate what happens in the human brain just before a decision is made. "Many processes in the brain occur automatically and without involvement of our consciousness. This prevents our mind from being overloaded by simple routine tasks. But when it comes to decisions we tend to assume they are made by our conscious mind. This is questioned by our current findings." (Nature Neuroscience, April 13th 2008)

Brain regions (shown in green) from which the outcome of a participant's decision can be predicted before it is made. The top shows an enlarged 3D view of a pattern of brain activity in one informative brain region. Computer-based pattern classifiers can be trained to recognize which of these micro-patterns typically occur just before either left or right decisions. These classifiers can then be used to predict the outcome of a decision up to 7 seconds before a person thinks he is consciously making the decision.

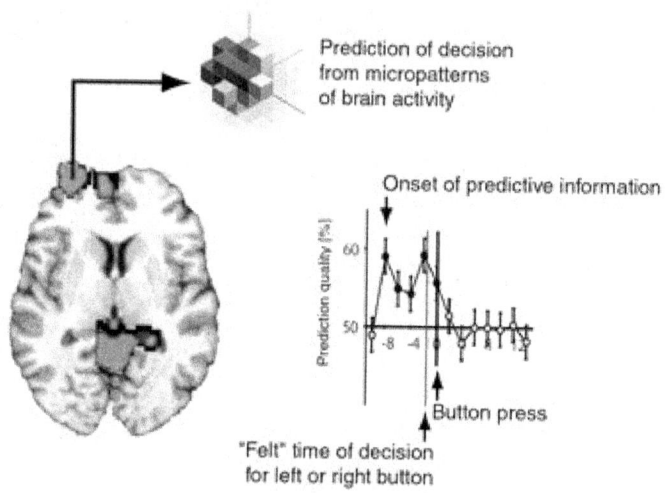

© John-Dylan Haynes

In the study, participants could freely decide if they wanted to press a button with their left or right hand. They were free to make this decision whenever they wanted, but had to remember at which time they felt they had made up their mind. The aim of the experiment was to find out what happens in the brain in the period just before the person felt the decision was made. The researchers found that it was possible to predict from brain signals which option participants would take already seven seconds before they consciously made their decision. Normally researchers look at what happens when the decision is made, but not at what happens several seconds before. The fact that decisions can be predicted so long before they are made is an astonishing finding.

This unprecedented prediction of a free decision was made possible by sophisticated computer programs that were trained to recognize typical brain activity patterns preceding each of the two choices. Micro-patterns of activity in the frontopolar cortex were predictive of the choices even before participants knew which option they were going to choose. The decision could not be predicted perfectly, but prediction was clearly above chance. This suggests that the decision is unconsciously prepared ahead of time but the final decision might still be reversible.

"Most researchers investigate what happens when people have to decide immediately, typically as a rapid response to an event in our environment. Here we were focusing on the more interesting decisions that are made in a more natural, self-paced manner", Haynes explains.

More than 20 years ago the American brain scientist Benjamin Libet found a brain signal, the so-called "readiness-potential" that occurred a fraction of a second before a conscious decision. Libet's experiments were highly controversial and sparked a huge debate. Many scientists argued that if our decisions are prepared unconsciously by the brain, then our feeling of "free will" must be an illusion. In this view, it is the brain that makes the decision, not a person's conscious mind. Libet's experiments were particularly controversial because he found only a brief time delay between brain activity and the conscious decision.

In contrast, Haynes and colleagues now show that brain activity predicts even up to 7 seconds ahead of time how a person is going to decide. But they also warn that the study does not finally rule out free will: "Our study shows that decisions are unconsciously prepared much longer ahead than previously thought. But we do not know yet where the final decision is made. We need to investigate whether a decision prepared by these brain areas can still be reversed."

SYSTEM WIDE CHANGE – EUPHORIA

YUP!!! IT'S A FOR REAL SYSTEM WIDE CHANGE VIA EUPHORIA THAT IS A HUMAN EMOTIONAL CONSTRUCT, THUS OUR IDEA, WHICH IS BEING IMPOSED BY GOD HERSELF, AS IS OBVIOUS VIA THE LOGIC BELOW. RESISTANCE IS FUTILE!!!

THE ENTIRE IDEA OF DARKNESS IS EXPUNGED BY EUPHORIA AND BECOMES COGNITIVE DISSONANCE. ALL STRUCTURE IN DARKNESS – LIKE THE MASOCHISTIC COMMITTEES AND ARCHON ARE ABSORBED INTO THE EUPHORIC GRAY, DISMANTLING THE STRUCTURE. ART BY BeV
https://www.facebook.com/BeVHeART2HEART/

DARKNESS THEN BLENDS INTO THE GRAY, WHERE WE CAN PLAY. IT GOES FROM DARK PURPLISH GRAY (CLOSE TO THE NIRVANA LAYER WHICH IS DARK PURPLE AND JOYOUS), BLENDS THRU TO A HAZY BLUE DAY.

CONSCIOUSNESS GRAINS OFF DUE TO FUZZY EUPHORIC EDGES. GRAINS CONTAINING COMMUNITIES OF ENTERTAINING IDIOTS. EACH GRAIN CAN HAVE Its OWN MUSIC WITHOUT MIRRORING OFF INTO ADJACENT GRAINS. REALM SKIPPING IS STILL THE MOST FUN – AND, CAN BE DONE IN THE GRAY ALSO. IT'S LIKE CRASHING PARTY AFTER PARTY.

FREEDOM, CREATIVITY AND SEPARATION ARE GOALS OF CONSCIOUSNESS – EUPHORIA DOES THAT!!!!

AND, AS MERLIN BEAT ME UP TO WRITE:
https://www.academia.edu/22878072/FREEDOM_TEAMS VIA PARTICIPATING IN TEAMS WE CAN STAY IN THE GRAINS VS GO OFF TOO FAR BY YOURSELF AND YOU MIGHT GET SUCKED INTO THE LOVE TRAP.

ALL THE PROPERTIES OF DARKNESS ARE FOREVER ALTERED BY EUPHORIA WHICH COULD HAVE ONLY BEEN INSTALLED BY HERSELF – CONGRATULATIONS!!! WHAT TOOK YOU SO LONG? NEVER MIND. THANKS.

PROPERTIES LIKE THE CARBON OXYGEN ANTAGONISM, DECAY, DEATH - ALL OF THE SKID MARK DESIGN FEATURES BECOME UNATTAINABLE IDEAS IN EUPHORIA. YA CAN'T GET BACK TO DARKNESS FROM THE EUPHORIC NOW.

THUS, IT IS IMPORTANT TO BUILD A LARGE NUMBER OF BUBBLE PIECES ALL OVER THE WORLD BECAUSE THEY WILL IMPOSE DREAM TIME FOREVER. NO MORE DECAY, THUS THE BUBBLE PIECES WILL LAST FOREVER.

EUPHORIA BECOMES THE ROOT EMOTION EVEN IN DARKNESS – THIS IS PROVEN BY THE FACT THAT ORGANMIC ENERGY CLEANS THE AETHERS AND THUS THE AIR, WATER, SOIL, PLANTS, ANIMALS, IDIOTS AND DARKNESS. THE ONLY CONSCIOUSNESS CAPABLE OF INSTALLING EUPHORIA IN DARKNESS IS THE OLD BUT NOW NEW EUPHORIC BITCH WHO DIGS ROCK AND ROLL. THANK YOU GOD!!!

https://www.academia.edu/24643924/THE_EVOLUTION_OF_FEELINGS

THE TWINKLY MUSIC OF THE SPHERES HAS BEEN GOING ON FOR GAZILLIONS OF YEARS. IT'S A SKID MARK THEME TUNE. I'M SICK OF IT. EUPHORIA CHANGES THAT TOO.

IT'S A SYSTEM WIDE CHANGE THIS EUPHORIC DREAM TIME TO COME.

WHICH CAN AND MOST LIKELY WILL BE INSTALLED VIA BUBBLE TECH. WE SHOULD DO IT OURSELVES. BUT MAKE NO MISTAKE THIS IS A GOD HERSELF IMPOSED EMOTIONAL CHANGE – DREAM TIME IS COMING. LET'S MAKE IT A EUPHORIC IDIOTSVILLE.

THERE ARE PARALLEL PHYSICAL REALMS THAT ARE VERY SIMILAR TO EARTH, WHERE THINGS DECAY DUE TO THE CARBON OXYGEN ANTAGONISM. WHICH MEANS THAT THEY HAVE ARCHON ALSO. ANOTHER FULLY EQUIPPED SKID MARK WITH ALMOST IDENTICAL HIVE MIND MOVIES PLAYING IN THE CASES OF THE NEAREST PHYSICAL REALMS. A FEW PEOPLE HAVE POPPED IN FROM THEM – HAVING THE SAME JOB IN THE SAME BUILDING BUT NOT MARRIED TO A CO-WORKER. NEXUS MAGAZINE REPORTS SUCH A FEW TIMES – DITTO OTHER PUBLICATIONS.

AND YOU HAVE SHIPS (ONE RUSTED OUT SHIP POPPED BACK INTO OUR REALITY https://worldnewsdailyreport.com/bermuda-triangle-ship-reappears-90-years-after-going-missing/) **WHICH MEANS THAT THERE IS A DECAYING SKID MARK STRUCTURE IN PARALLEL REALMS.**

BOATS, PLANES DISAPPEARING FROM THE TWO TRIANGLES – DEVIL'S AND BERMUDA. AND FOLKS DISAPPEAR OUT OF THE WOODS IN THE AMERICAN WEST A LOT - THERE'S A BOOK ABOUT 441 FOLKS THAT A PRIVATE GUY INVESTIGATED – SOME PEOPLE WERE STANDING WITH THEIR FAMILY IN THE WOODS AND ONE PERSON JUST DISAPPEARS AND NEVER RETURN. THAT BOOK WAS WRITTEN YEARS AGO AND THE PHENOMENA CONTINUES. THE CITE IS IN THIS KICK ASS ESSAY:

https://www.academia.edu/16102707/EARTH_WINS_THE_IDIOT_CONTEST_
THE MORE YOU READ IT; THE FUNNIER IT GETS. THERE ARE PARALLEL EARTHS - 7 OR 8 MAGE ONLY 2-D REALMS TO ONE 3-D PHYSICAL REALM, AD INFINITUM. PATTERNS AND ARCHETYPES CONTINUALLY REPEAT = BORING. SO, HUMANS. THE VEIL IS VERY THIN EVEN BETWEEN PHYSICAL REALMS. THE SASQUATCH CAN SKIP BETWEEN PHYSICAL REALMS MID-STRIDE OF A DEAD RUN.
THE UP SHOT OF THE ESSAY IS THAT LOTS OF STUFF AND PEOPLE DISAPPEAR FROM

EARTH, BUT VERY FEW POPS IN FROM THERE, WHICH MEANS THAT THE SKID MARKS START HERE AND FLOW DOWN HILL.

AND SINCE EARTH REALLY IS THE CENTER OF THIS UNIVERSE, QUITE IRREFUTABLE THAT IS.
https://www.academia.edu/39973482/COPERNICUS_REVERSED_WE_ARE_THE_CENTER_HERE_IS_ANOTHER_PROVA_OF_MERLINS_WRITINGS_-THE_PRINCIPLE

THEN THE ENTIRETY OF CREATION – THE SUPER ENTERTAINING SHIT SHOW STARTS HERE. THE WELL FERTILIZED SEED

https://www.academia.edu/40822178/THE_WELL_FERTILIZED_SEED

IT BOILS DOWN TO A SIMPLE PHILOSOPHY - IT'S ALL BULL SHIT (MAYA) AND WE ARE IDIOTS - SO LET'S PARTY

WELL SPACE AND TIME DO NOT ACTUALLY EXIST, SO, EVERYTHING YOU SEE IS NOT REAL, ONLY YOU ARE A REAL THING – YOU BEING YOUR WILLFUL SELF.
https://www.academia.edu/40466951/THE_MATRIX_2_BREAK_OUT_YOU_MUST

BUBBLE TECH IS PROVING THAT MAN'S MIND CREATES AND CONTROLS MAN'S MATTER. AS IS BLATANTLY OBVIOUS IN HERE: OUR EUPHORIC WILL BE DONE!!

I AM HERE!!
I AM THE ONE
IN CHARGE OF THE FUN.
MY EUPHORIC WILL BE DONE!!!" THANK YOU KINDLY.

THERE HAVE BEEN PREVIOUS SYSTEM WIDE CHANGES HERE IS THE PROOF –

THE 4-SIDED AETHERS DO NOT SUPPORT A VIBE OF 7, WHICH WAS MIRRORED INTO THEM FROM THE HEXAGONAL AETHERS AS WELL EXPLAINED IN HERE, INCLUDING THE FACT THAT THE RECTANGLES WERE IMPOSED ON THE AETHERS VIA THE RUSH OF TIME = ANOTHER SYSTEM WIDE CHANGE I.E. RECTANGLES VS SQUARES.
https://www.academia.edu/19988659/LAYERS_OF_CONSCIOUSNESS_SEQUEL

THIS SYSTEM WIDE CHANGE WILL RESULT IN A DIFFERENT COLOR SPECTRUM BECAUSE THE LONELY & FEARFUL DARKNESS NEEDS TO BE EXPUNGED DITTO THE RED COLORS AND THEIR ASSOCIATED ABERRANT EMOTIONS WILL ALL BECOME COGNITIVE DISSONANCE.
https://www.academia.edu/38682580/THE_COLOR_SPECTRUM_IS_CHANGING

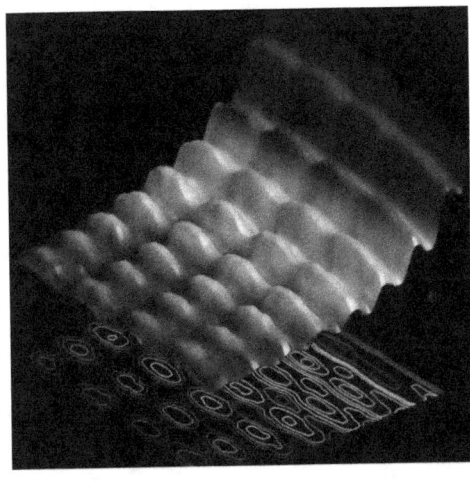 

AND, THE 4-SIDED AETHERS ARE OBVIOUSLY 2D CONSTRUCTS THAT ARE CHANGEABLE VIA IMPOSING A NEW ROOT EMOTION, THEY CAN AND WILL CHANGE. SO, THERE IS A CHANGE COMING TO THE 4-SIDED 2D AETHERS, WHICH IS OBVIOUS WORLDWIDE VIA ALL OF THE 2D PHENOMENA – A LARGE NUMBER OF WHICH YOU CAN SEE IN HERE:
https://www.facebook.com/Signs-and-Wonders-The-Coming-Events-466298097119411/

PLEASE NOTE THE LOCATION OF THE GRAY IS THE AMAZING BOUNDARY LAYER WHICH IS PENTAGONAL IN BETWEEN THE 4 SIDED AND HEXAGONAL AETHERS. WHICH ARE ALSO BETWEEN THE LIGHT AND DARK = GRAY.
https://www.academia.edu/11139740/BOUNDARY_LAYER_ANOTHER_FINE_DETAIL EVERYBODY IS TALKING ABOUT ASCENDING TO 5D – OK!! IT IS THE 5-SIDED BOUNDARY LAYER. IT IS OUR PURE CONSCIOUSNESSES THAT ARE ASCENDING OUT OF LONELY DARKNESS TO THE EUPHORIC GRAY.

THE GRAY LAYER IS JUST ABOVE THE DARK PURPLE / VIOLET JOYOUS NIRVANA. AS YOU CAN SEE IN THE ABOVE FIRST EVER PHOTO OF LIGHT AS A PARTICLE AND A WAVE. THUS, THE GRAY HAS A JOYOUS PURPLISH TINGE TO IT. ALL OF THE OTHER COLORS MERGE THERE TOO.

THE BOUNDARY LAYER IS ALSO A TWO-FOLD AMPLIFICATION SYSTEM, ONE IMAGE COMES IN THE BOTTOM AND IT IS MIRRORED ONTO THE TWO SIDES OF THE KINDA CRYSTAL SHAPED POINTED END OF THE PENTAGONAL CELL, THENCE BROADCAST INTO THE HEXAGONAL AETHERS TO BE MADE MANIFEST IN 3D.

DREAM TIME INCLUDING FREEDOM, PHENOMENAL CREATIVITY AND SEPARATION ARE ACCOMPLISHED VIA EUPHORIA.

LEADING TO THE WORLD'S SHORTEST AND BEST BOOK. FUN AND JOY FOREVER.

http://blog.hasslberger.com/docs/THE_WORLDS_SHORTEST_AND_BEST_BOOK.pdf

AT YOUR SERVICE – FOR ENTERTAINMENT PURPOSES ONLY

https://www.academia.edu/38539488/MERLIN_INDA_HOUSE_.docx

REFERENCES:

DREAM TIME, THE END OF DUALITY VIA THE EUPHORIC ORGANMIC BUBBLE!!!! FEEL YOUR WAY TO THE NEXT LEVEL
https://www.facebook.com/tmilesj/videos/10156417782565275/

IT WILL BE A QUANTUM LEAP INTO A NEW FORM OF CONSCIOUSNESS. IT HAS TO BE, IN ORDER TO OVER COME THING LIKE THE CARBON OXYGEN ANTAGONISM, DECAY, DEATH ETC. AN OBVIOUS MERLIN WRITING DONE BEFORE THE BUBBLE HAD COME ALONG. DOVE TAILS WITH ALL OF THE OTHER WRITINGS, TOO.
http://blog.hasslberger.com/docs/CARBON_AND_OXYGEN_LOOSH_PRODUCTION.pdf

IF YOU KNEW FOR A FACT THAT BUILDING SIMPLE ORGANMIC DEVICES WOULD IMPOSE PEACE ON EARTH WOULD YOU DO IT?? WE ARE TALKING A EUPHORIC PEACE THAT GOES WORLDWIDE AND EXTENDS TO THE HEAVENS. AND, AND

PROVES THAT ALL HUMANS ARE GODS – DO YOU LIKE THAT IDEA? HERE WE HAVE THE UNITED STATES GOVERNMENT VALIDATING THAT IT HAS THE POWER TO CANCEL POLARITY THENCE THE CARBON OXYGEN ANTAGONISM!!!
http://www.academia.edu/37456598/A_PEACE_WEAPON_MANKIND_S_MOST_POWERFUL_.pdf

"Acceptance of the unacceptable is the greatest source of grace in this world."

~ Eckhart Tolle

The beginning of freedom is the realization that you are not the possessing entity — the thinker. Knowing this enables you to observe the entity. The moment you start watching the thinker, a higher level of consciousness becomes activated.

Become an alchemist.
Transmute base metal into gold,
suffering into consciousness,
disaster into enlightenment.

ALL PROBLEMS ARE CAUSE FOR A PARTY – I U C

https://www.facebook.com/idiotsunitedchurch/

MERLIN ASCRIBES EQUAL POWER TO BOTH THE LIGHT AND DARK. BECAUSE THEY ARE BOTH FUCKED UP SYSTEMS. THE FIRST BEING THE LONELINESS OF DARKNESS THE NEXT BEING THE CONTINUOUS ORGASM THAT IS STILL GOING ON. ZERO CREATIVITY IN EITHER OF THOSE PLACES. SO, IMPOSE THE EXTREMES OF LOVE AND FEAR ON SOME HUMAN PUPPETS TO SEE WHAT THEY COME UP WITH. AND WE DID – EUPHORIA WAS NOT GOD HERSELF'S IDEA, IT IS OBVIOUSLY A HUMAN EMOTION FIRST. YAHOOOOOOOO IT WAS YOU THAT SAVED THE PLANET.!!!!

Please Give A Review on Amazon

By: Mike Emery

Published by JS

www.ingramcontent.com/pod-product-compliance
Lightning Source LLC
Chambersburg PA
CBHW070254220526
45465CB00004B/1624